IMAGES
of America

AROUND CONNELLSVILLE

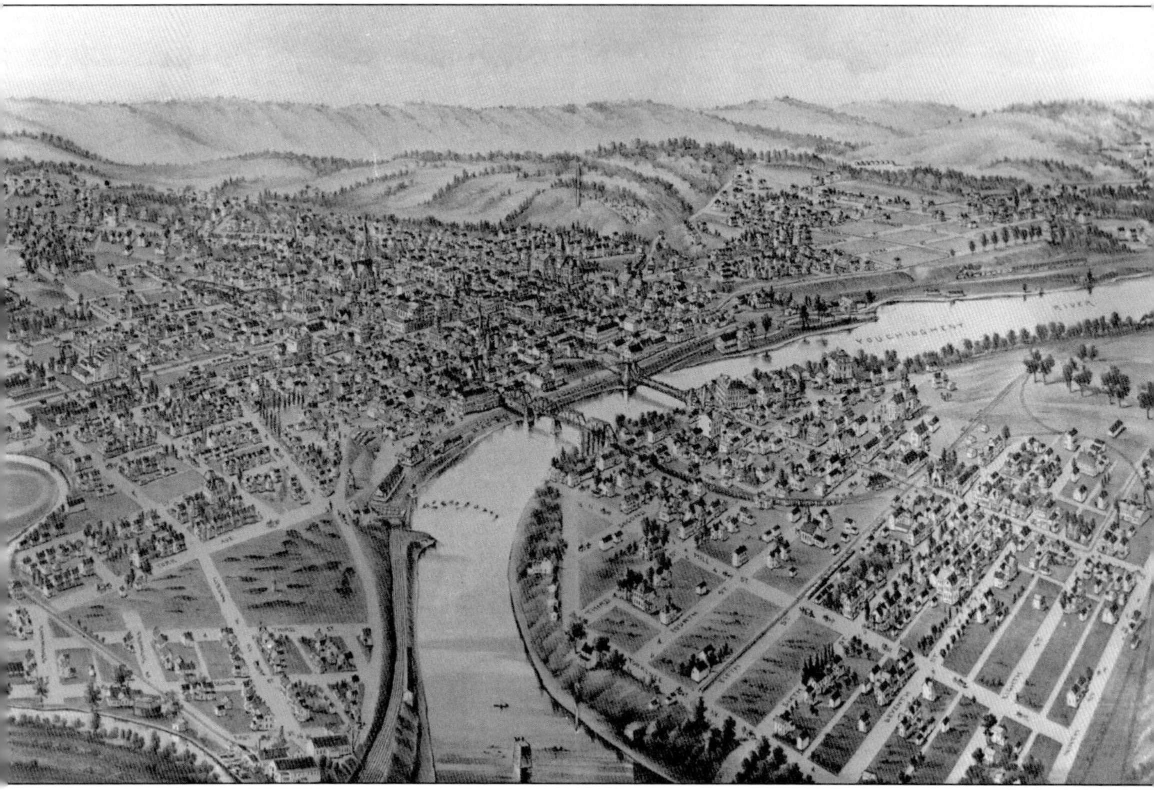

This 1897 map shows the Youghiogheny River separating Connellsville, on the left, from New Haven, on the right. Prominent features clustered along the river include the Pennsylvania Railroad Bridge in midstream, followed by the old suspension bridge linking the commercial center on Crawford Avenue in Connellsville with its extension on the west side. In addition to many notable landmarks, the map reflects a thriving community nine years before its centennial celebration. (Courtesy of the Connellsville Area Historical Society.)

On the Cover: Members of the Molinaro Band pose at St. Rita's Grotto in 1948. The group was formed in 1913 by Michael Molinaro, who served as director until 1936. He was succeeded by his brother Amedeo (first row, far right). Carmine Molinaro Sr., standing next to Amedeo, also served as director. In 2013, the band celebrated 100 years of making music around Connellsville. (Courtesy of Henry Molinaro.)

IMAGES
of America

AROUND CONNELLSVILLE

Paul E. Eckman, Karen Hechler, and the
Connellsville Area Historical Society

ARCADIA
PUBLISHING

ISBN 978-0-7385-9935-9

Published by Arcadia Publishing
Charleston, South Carolina

Printed in the United States of America

Library of Congress Control Number: 2012955524

For all general information, please contact Arcadia Publishing:
Telephone 843-853-2070
Fax 843-853-0044
E-mail sales@arcadiapublishing.com
For customer service and orders:
Toll-Free 1-888-313-2665

Visit us on the Internet at www.arcadiapublishing.com

*To Paula Salatino Beucher, Marjorie Eckman,
Van Hughes, Bill Sechler, and, in memory, Bill Balsley*

CONTENTS

ACKNOWLEDGMENTS

We are deeply grateful to the following individuals for assistance, information and pictures: Roxanne Abramowitz, Donna Campbell Allen, Marilyn Banks, Jessica Kadie Barclay, Howard Barnhart, David Baughman, Jeanne Baughman, Tillie Berg, Paula Salatino Beucher, Barbara Bielecki, Pam Braxton, Robin Bubarth, Bob Buttermore, Barbara Crossland Chess, Dan Cocks, Ed Cope, John Ed Dailey, Ted Davis, Barbara Eckman, Joseph Eckman, Marjorie Eckman, Michael Edwards, Joan Emanuel, Thelma Evans, Betty Flynn, Denny Gaal, Barbara Gallagher, Ginger Crossland Glisan, Elsie Haley, Harold Hartz, Van Hughes, John Husband, Janice Jaynes, Ken Jaynes, Sr., Alan Jones, Colivia Jones, Judy Keller, Barbara Keefer, Tony Keefer, Mary Ann Komazec, Tom Korba, Joe Kurtz, Robin Leighty, Arlene Leonard, Pete Maricondi, Clyde Martz, Joseph McKitrick, Carmine Molinaro, Jr., Henry Molinaro, Margaret Molinaro, Fotenie Mongell, Deborah Mullaney, Ann Louise Natale, Anita Nicholson, Tinkey Nist, Bob Omatick, Paul Paterra, Bob Percy, Harry Porter, Kay Porterfield, Felix Prestia, Dexton Reed, Judy Reed, Susan Rudnik, Tom Rusnack, Sandy Russell, Evan Sanders, Alan Sandusky, Albert Sandusky, Millie Sandusky, Marion Scardina, John "Wally" Schroyer, Beverly Shaffer, Casey Sirochman, Patrick Stefano, Mary Lou Sullivan, Linda Suter, Bert Swartzwalder, Laura Szepesi, Oscar Tissue, and Tom Zwierzelewski.

Special recognition is extended to Bill Sechler whose technical skills and general knowledge of Connellsville facilitated the timely completion of this volume.

We are also thankful to these organizations and businesses that provided assistance: Bud Murphy's, Connellsville Area Historical Society, *Connellsville Memories*, Connellsville Public Library, Connellsville Senior Tigers, *Daily Courier*, Fayette County Cultural Trust, Girl Scout Little House Society, Molinaro Law Offices, New Haven Hose Company, Scottdale Bank and Trust Company, *Scottdale Independent Observer*, Veterans of Foreign Wars Post 21, West Overton Museums, and Widmer Engineering.

Unless otherwise noted, all images are from the collection of the authors and the Connellsville Area Historical Society.

Due to publishing guidelines, we could not include all available photographs in this volume; however, they will be used in future projects sponsored by the Connellsville Area Historical Society.

Introduction

Connellsville is a tale of two frontier settlements divided by a wild and scenic river. Located on the east bank of the Youghiogheny River, Connellsville was incorporated in 1806. It was named for Zachariah Connell, who came to the area from Virginia around 1772. He surveyed the land, planned the town, and functioned as a land agent for the families of Gov. Robert Dinwiddie of Virginia and Benjamin Chew in Philadelphia. The first known English settler, however, was William McCormick, who migrated from Winchester, Virginia, in 1770. Other pioneers included Joseph Page, David Barnes, Benjamin Wells, George Mathiot, Daniel Rogers, Peter Stillwagon, John Gibson, and Samuel and Caleb Trevor.

The west side of the river was settled in 1753 by William Stewart, who was forced to leave the area when threats from Native Americans increased after George Washington surrendered Fort Necessity in 1754. In the following year, Gen. Edward Braddock led his army through Connellsville on his ill-fated campaign to oust the French from the forks of the Ohio River. For two days, on June 28 and 29, 1755, General Braddock camped on the west side of the Youghiogheny River. On June 30, the army forded the river at what was known as Stewart's Crossing and bivouacked on the east bank before marching northwest to Mount Pleasant.

The first permanent settler on the west side was William Crawford, who came from Virginia in 1765. He was a personal friend and business associate of George Washington. During the American Revolution, Crawford recruited a battalion that was part of the 7th Virginia Regiment. They fought in battles on Long Island and at Trenton, Princeton, and Brandywine Creek. In 1782, Crawford led an expedition against Native Americans near the Sandusky River in Ohio, where he was captured and killed. His land on the west side was eventually sold to Isaac Meason, who laid out the town of New Haven. The town was chartered in 1839 but not formally organized until 1867. After local referendums, New Haven merged with its east side neighbor in 1909, thereby making Connellsville the first city in Fayette County.

Connellsville was always a manufacturing and transportation center. After crossing the Allegheny Mountains, pioneers stopped at the settlement to build boats before descending the Youghiogheny River to Pittsburgh. In the late 1700s, there were several iron furnaces and foundries in or near Connellsville. They included the Etna furnace, operated by the Gibson family, and the Union furnace, which was built in Dunbar by Isaac Meason. Reportedly, Meason experimented with coke as an alternative to charcoal for smelting iron, but the attempt did not produce an acceptable furnace fuel. The first successful production of coke, using a beehive oven, occurred in Connellsville in 1833, when a Mr. Nichols constructed an oven for Leroy Norton, who used the fuel in his foundry. These efforts marked the beginning of a new industry that transformed the region.

Most of the coal mines and coke ovens were concentrated above the Pittsburgh coal seam, an area 50 miles long and approximately 5 miles wide along the base of the Chestnut Ridge, from Uniontown to Greensburg. Since Connellsville was near the geographical center, the area was called the Connellsville Coke Region.

By 1880, the area had 7,000 ovens and produced two-thirds of the nation's supply of coke. A census of the industry during World War I revealed the presence of 175 coke works and approximately 35,000 beehive ovens. Most of the operations were eventually owned or leased by the Henry Clay (H.C.) Frick Coke Company, which became a subsidiary of US Steel in 1901.

The mining districts were not free of conflict. One of the most violent strikes occurred in 1894. During this labor action, Frick's chief engineer, John Paddock, who was also a member of the Connellsville Borough Council, was killed at the Davidson Mine. Nearly 100 strikers were arrested. Two foreign workers were convicted on questionable evidence and sentenced to 12 years in prison. There was other violence in and around Connellsville. On another occasion, rumors circulated that 400 armed men were marching to the city. Alarmed citizens rushed to the armory for weapons, but guards around the facility blocked their entrance. Labor conflict in the mining and coke industries continued until World War II.

Coke was "king" in Connellsville until the Great Depression. The extensive mineral reserves of the area attracted eight railroads to the region, including the Pittsburgh & Connellsville, the first line to reach the city, in 1855; the Baltimore & Ohio (B&O); Pennsylvania (PRR); Pittsburgh, McKeesport & Youghiogheny; Pittsburgh & Lake Erie (P&LE); Pittsburgh & West Virginia; Norfolk & Western; and the Western Maryland.

While the railroads linked Connellsville to the outside world, an extensive light rail system provided passenger and freight service within the region. In 1917, the West Penn Railway Company was incorporated, with its headquarters, car storage, and maintenance facilities in Connellsville. West Penn Electric was the parent company. It also operated a large power plant on the west side of the river, supplying electricity to customers in a four-county area.

About 100 years after the incorporation of Connellsville, manufacturing evolved from boatyards, iron furnaces, lumber, and cotton mills to industries that, in addition to coke, produced steam-driven automobiles, locomotives, chemicals, rolled steel, safes, heaters, metal coatings, bricks, glass, mine machinery, plumbing supplies, beer, and whiskey. The concentration of industry and transportation made Connellsville the financial and retail center of Fayette County until the opening of the lower Connellsville coalfield, which shifted new investment to Uniontown.

In the early 1900s, Connellsville had five banks, eight hotels, two department stores, seven clothing establishments, fourteen food markets, three furniture outlets, five pharmacies, and three hardware emporiums. Most of the businesses concentrated around Brimstone Corner, which remained an active retail and commercial center until the construction of an indoor mall in the 1970s along Route 119, which then became the preferred location for new businesses.

The wealth generated in Connellsville enabled residents to support a hospital, a public library, a modern school system, and an airport. In 1899, Andrew Carnegie donated $68,000 for a library, supplemented with local appropriations. A small hospital was established by the Commonwealth of Pennsylvania in 1891. Later, it became the Connellsville State Hospital. In 1938, Connellsville, in cooperation with Fayette County, constructed an airport in Dunbar Township that was used by the Army Air Corps during World War II.

Connellsville was a center of learning. The first high school was erected on Fairview Avenue in 1895. Another secondary school was established by Immaculate Conception Parish in 1921 and later became Geibel High School. In the mid-1960s, Pennsylvania State University offered college courses at Connellsville High School. This initiative was followed by the establishment of a campus in Fayette County.

Connellsville was and continues to be a place for celebration. On the last full weekend in June, citizens reenact Braddock's Youghiogheny River crossing. Parishioners at St. Rita's Roman Catholic Church remember their Italian heritage with a festival, and the whole city cheers participants in a five-kilometer run organized to commemorate John Woodruff's victory in the 1936 Berlin Olympics.

Connellsville was settled and initially prospered because of a river. Native Americans recognized its strategic location when they forged the Catawba Path through a wilderness that became a city. The river continues to shape the history of Connellsville. Today, a new generation of "pathfinders" hike or bike the Great Allegheny Passage through Connellsville, thereby creating a new gateway to the Laurel Highlands Recreational Area, which includes world-class resorts, state parks, historical sites, and the singular beauty of Frank Lloyd Wright's Fallingwater.

One

STARTING WITH THE RIVER

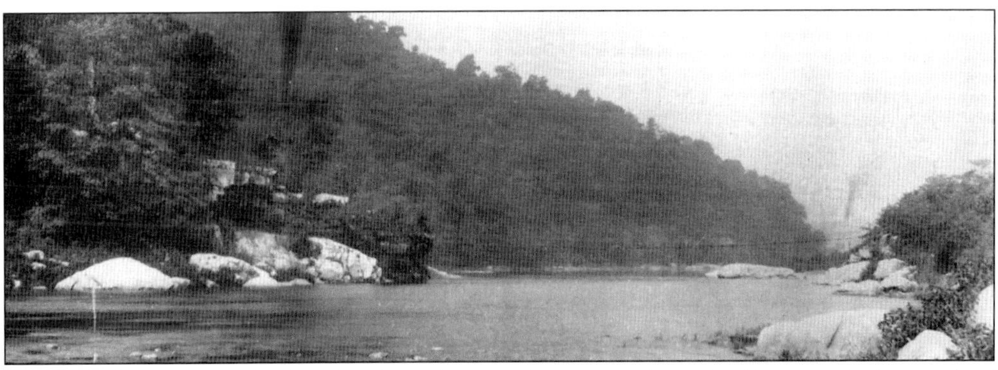

This is a 1909 view of the Youghiogheny River as it emerges from the Chestnut Ridge south of Connellsville. The river originates in the mountains of West Virginia and flows 132 miles in a northerly direction to its confluence with the Monongahela River at McKeesport. The name Youghiogheny has its origins in one or more Native American dialects. Most linguists agree that it means "twisting, turning, frothing water." (Courtesy of Felix Prestia.)

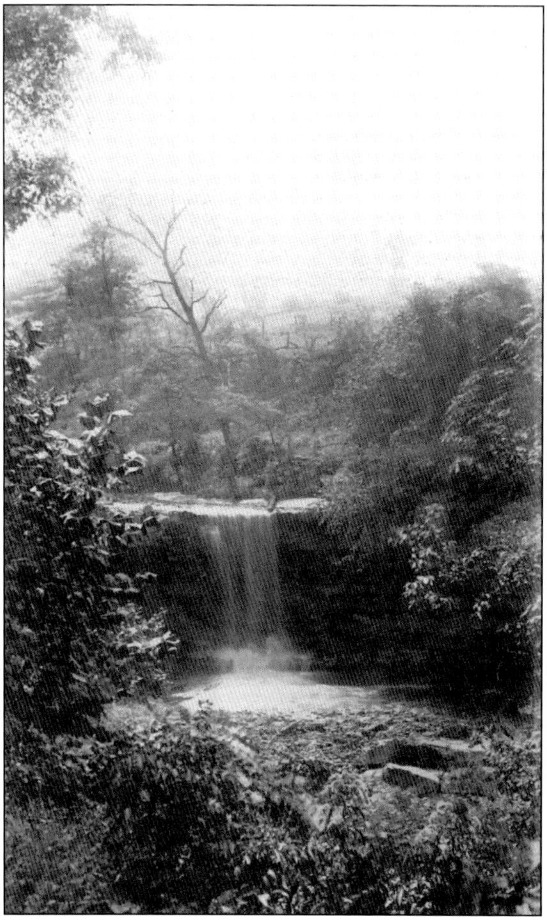

On June 28, 1755, Gen. Edward Braddock and his army left the Christopher Gist settlement in Dunbar Township for these falls on Opossum Creek in Connellsville. They stayed here for two days. The site was his 12th encampment after leaving Fort Cumberland, Maryland. The falls were subsequently named for John Robinson, who operated a mill and distillery on the property around 1818.

The postcard below depicts Stewart's Crossing, looking east across the Youghiogheny River around 1753. William Stewart operated a ferry service here until Native Americans forced settlers to leave after the French captured Fort Necessity. General Braddock forded the river here on June 30, 1755. After the crossing, he camped near the Davidson farm, in what is now Connellsville Township, before continuing his march to Fort Duquesne. (Courtesy of Sandy Russell.)

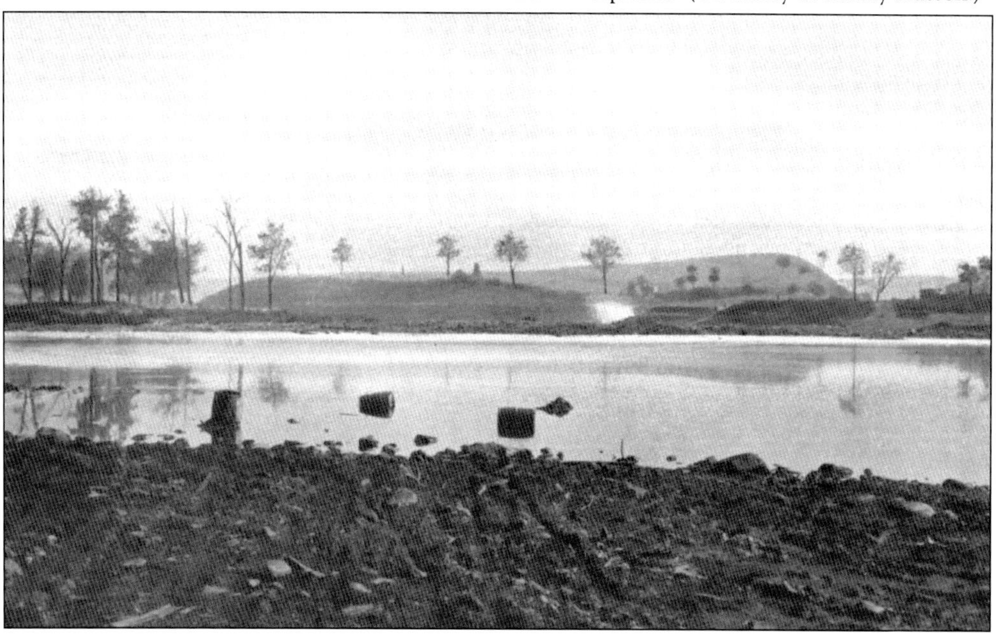

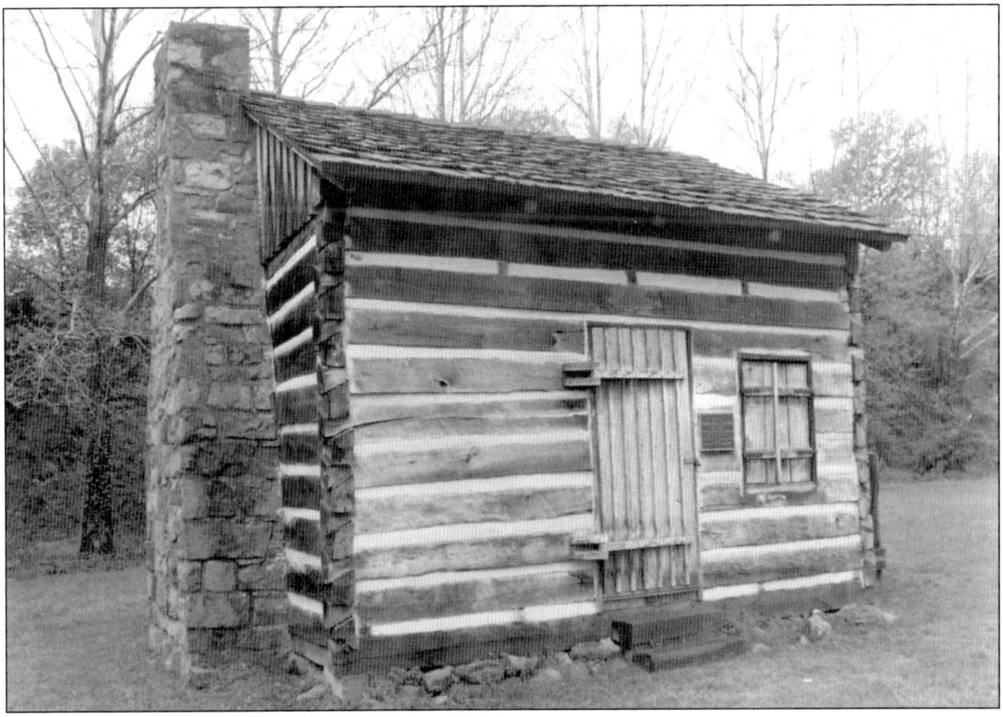

The replica of William Crawford's cabin adjacent to Yough River Park (above) was built by the Connellsville Area Historical Society in 1976. Crawford was a boyhood friend of George Washington and served in the 1758 Forbes Expedition, which ousted the French from the Ohio Valley. Crawford was impressed with the area and returned in 1765 to become the first permanent settler on the west side of the Youghiogheny River.

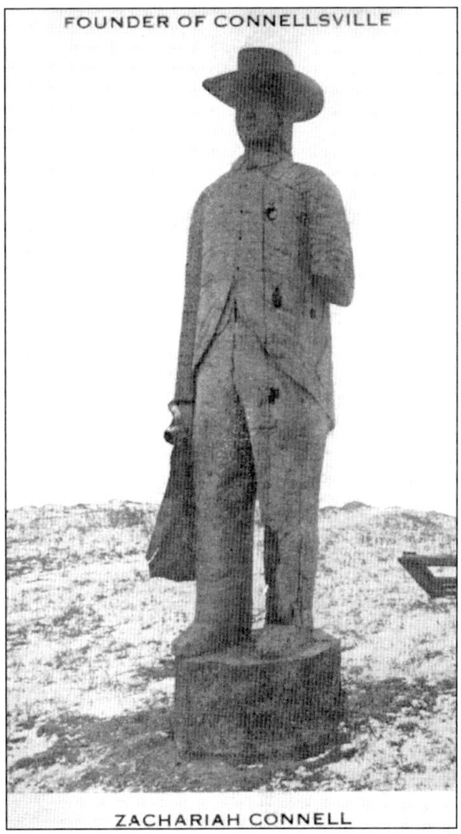

There are no public monuments dedicated to Zachariah Connell except this wooden statue, which was carved from a poplar log and mounted on the bell tower of the Union School around 1879. When the school was demolished, the statue was removed and subsequently disappeared. (Courtesy of Bert Swartzwalder.)

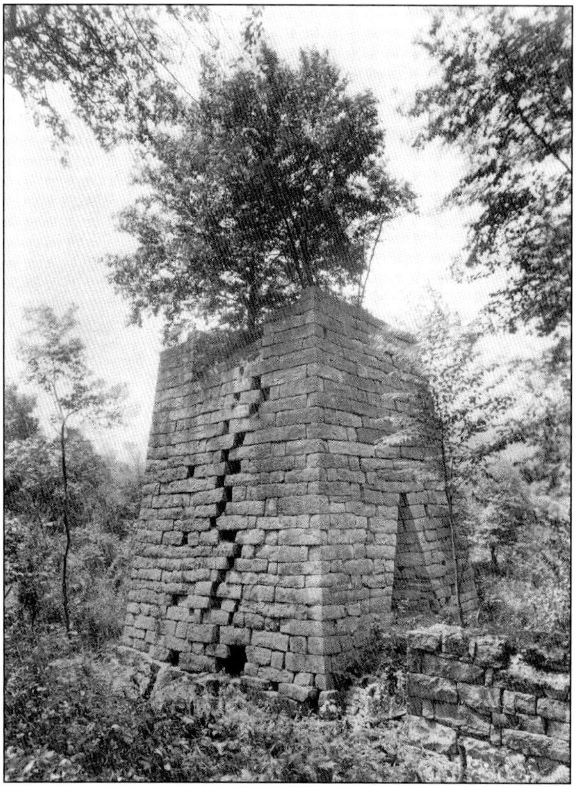

The smelting and casting of iron was a major enterprise in Fayette County, beginning with the Alliance furnace on Jacobs Creek, which was blown in 1789. Isaac Meason, who was associated with John Gibson, built the Union furnace two years later on Dunbar Creek. Gibson also established the Old Laurel furnace in Dunbar Township, which is seen here around 1920. (Courtesy of West Overton Museums.)

The sketch below is of the Gibson house on West Patterson Avenue. It was built by Joshua Gibson in 1870 and is one of the oldest structures in Connellsville. Joshua was the grandson of ironmaster John Gibson. John Gibson's sons, Thomas and Joseph, constructed the Etna furnace, which was near the present site of Falcon Stadium. It was the center of an iron-manufacturing village called Gibsonville. (Rendering by Donna Campbell Allen.)

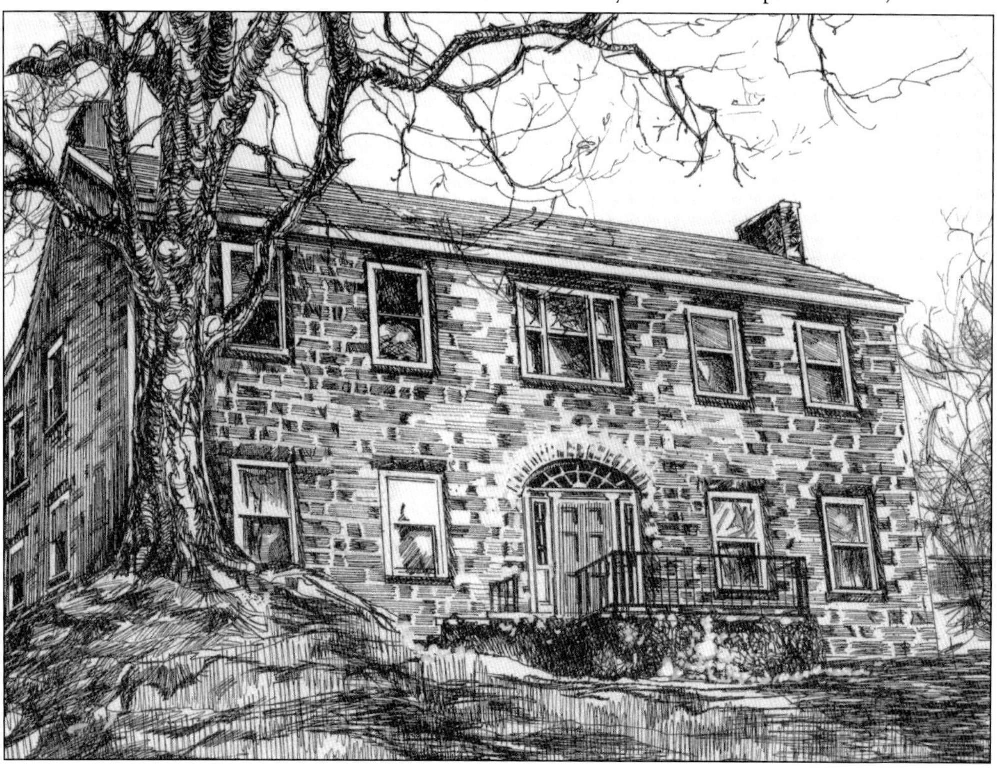

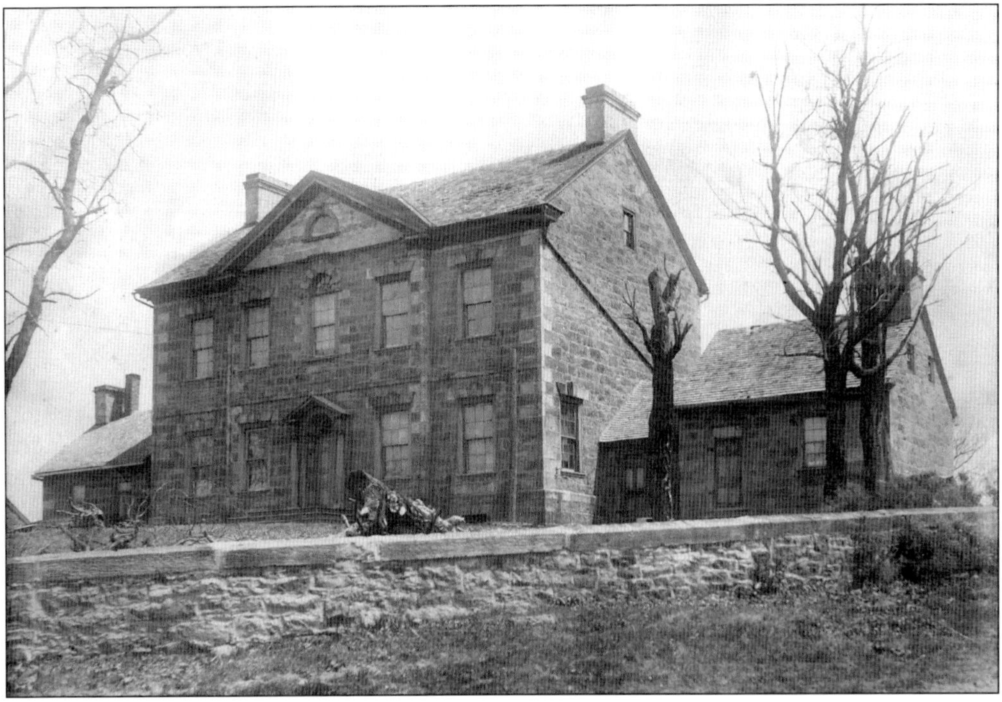

Mount Braddock, the home of Isaac Meason, is seen above in the early 1950s. Meason erected the limestone building in 1802 on land first settled by Christopher Gist in 1753. The mansion is recognized as the finest example of Georgian Palladian architecture west of the Allegheny Mountains. The property is privately owned and was designated a National Historic Landmark in 1990. (Courtesy of West Overton Museums.)

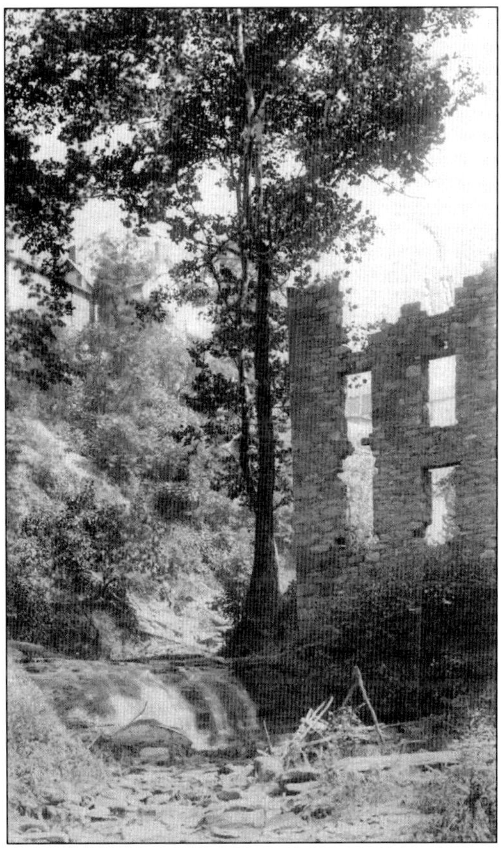

This wall, photographed in 1898, is all that remains of a cotton factory erected by Philo Norton in 1812. It was situated in Hogg's Hollow (East Park) and used the falls on McCoy Run to generate waterpower. The top floor of the four-story building was almost level with Fairview Avenue. A bridge reportedly connected the road to the top level of the factory.

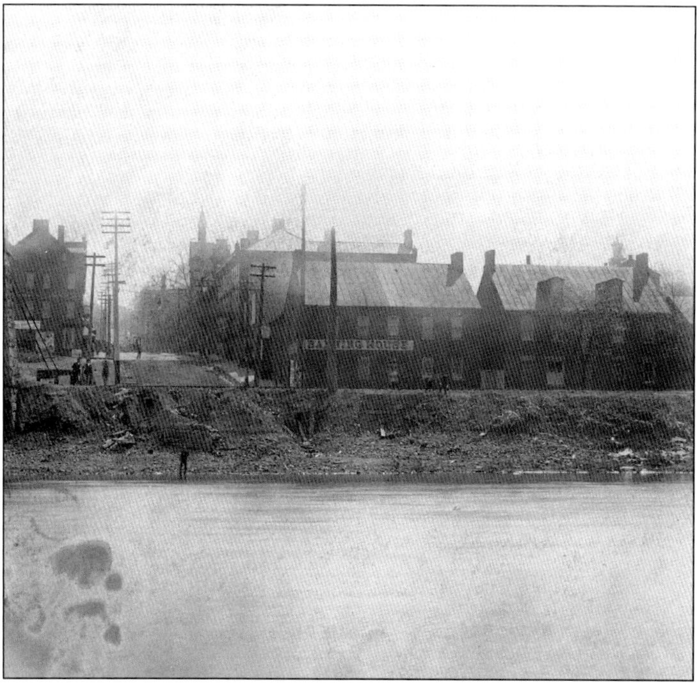

The intersection of Main (West Crawford Avenue) and Water Streets is pictured in the mid-1800s from the west side of the Youghiogheny River. The Banning House occupies the corner property on the right. It was a tavern operated by Anthony Banning. He was among the first 15 pioneers who purchased a lot directly from Zachariah Connell before 1806. Banning was an itinerant Methodist minister and served as a member of the first borough council.

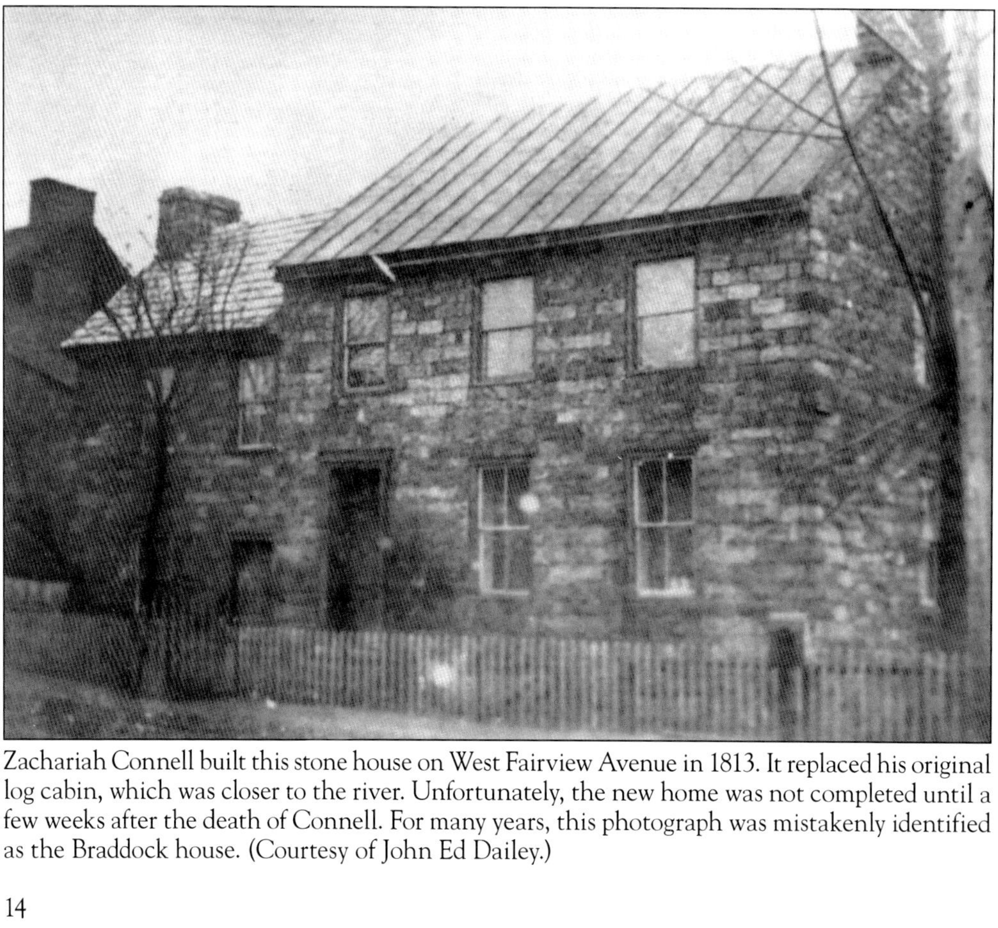

Zachariah Connell built this stone house on West Fairview Avenue in 1813. It replaced his original log cabin, which was closer to the river. Unfortunately, the new home was not completed until a few weeks after the death of Connell. For many years, this photograph was mistakenly identified as the Braddock house. (Courtesy of John Ed Dailey.)

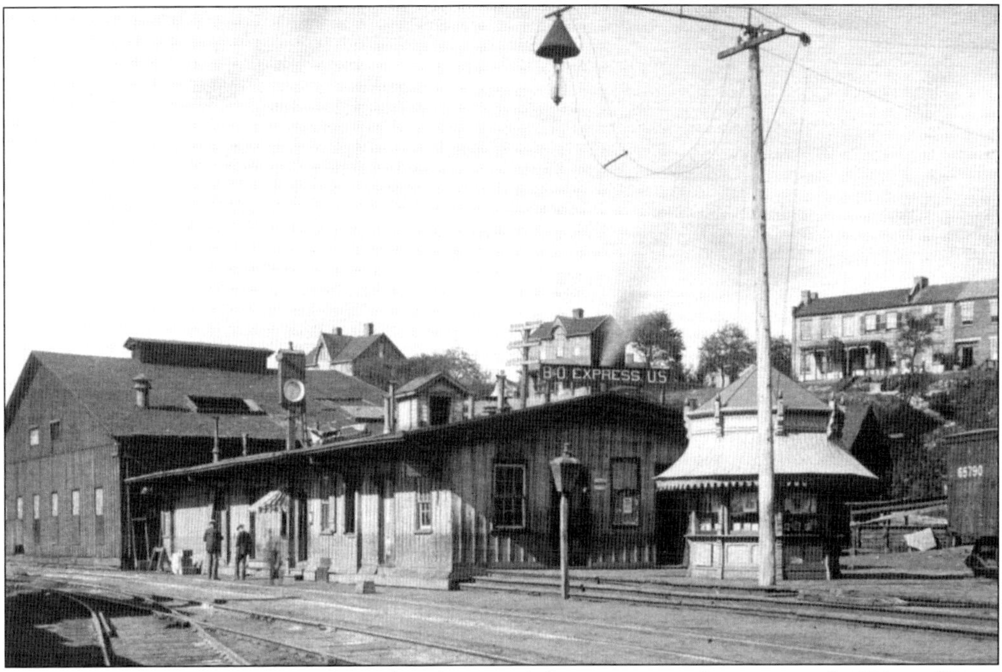

The Pittsburgh & Connellsville Railroad erected this passenger depot around 1870 on Water Street. The large building directly behind the station was a repair facility for locomotives and passenger cars. The facilities became B&O property in 1875 when the railroad entered into a long-term lease agreement with the Pittsburgh & Connellsville.

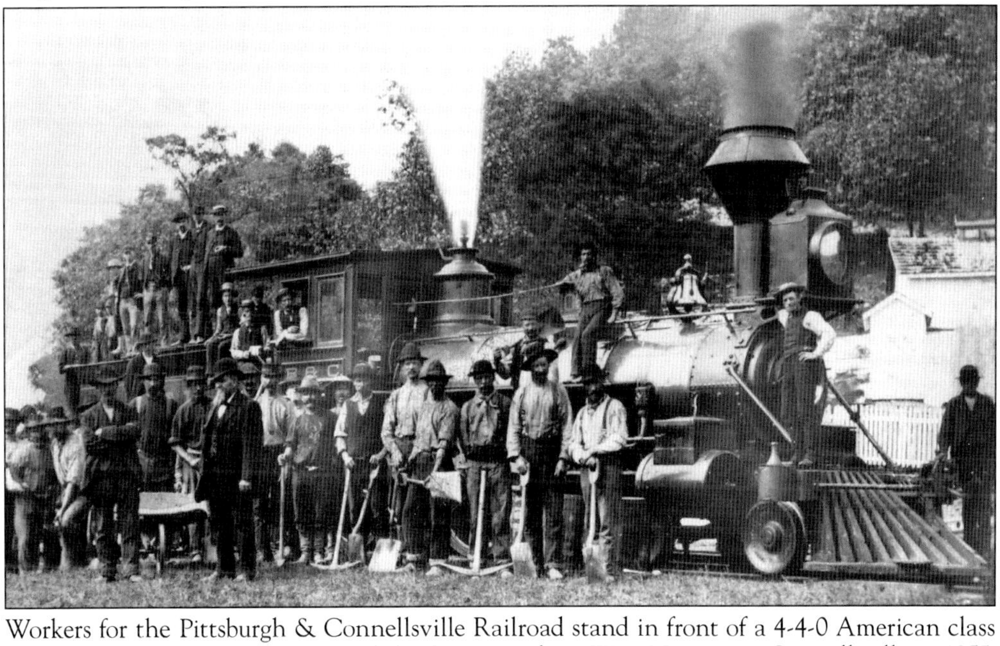

Workers for the Pittsburgh & Connellsville Railroad stand in front of a 4-4-0 American class locomotive that reportedly powered the first train from West Newton to Connellsville in 1855. Daniel Rogers Davidson is standing on the front of the engine, beneath the headlamp. Davidson was an early promoter of the railroad. Also, his farm north of Connellsville became the location of the Davidson Mine and Coke Works in 1856.

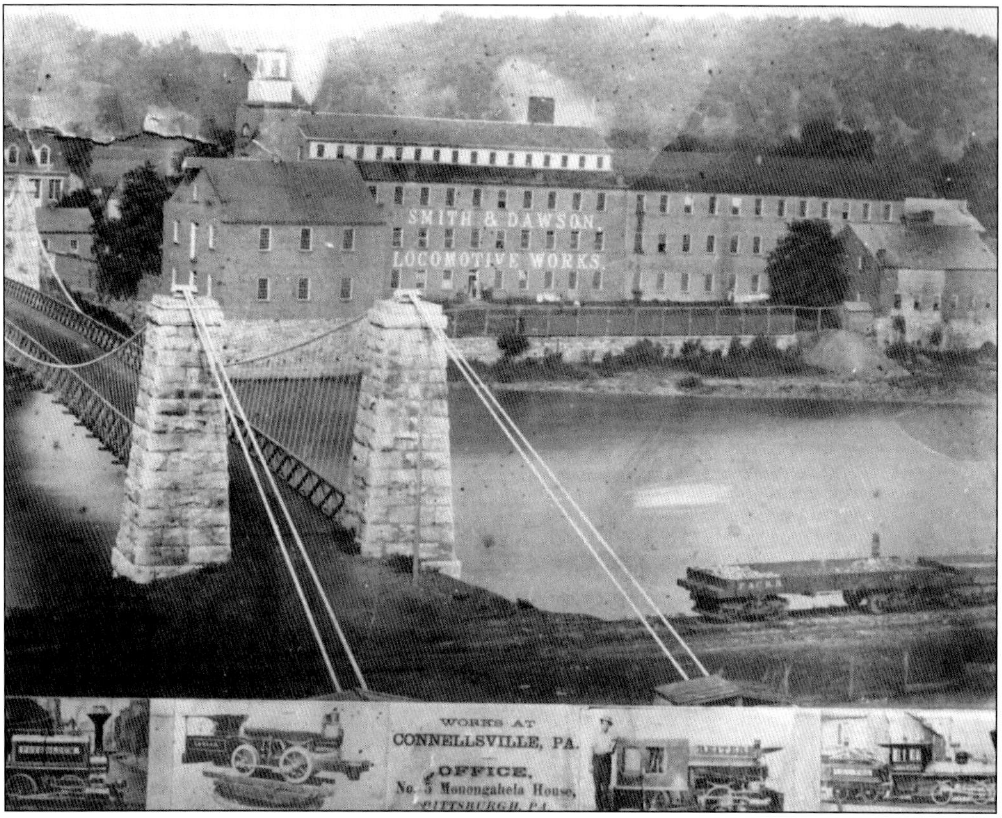

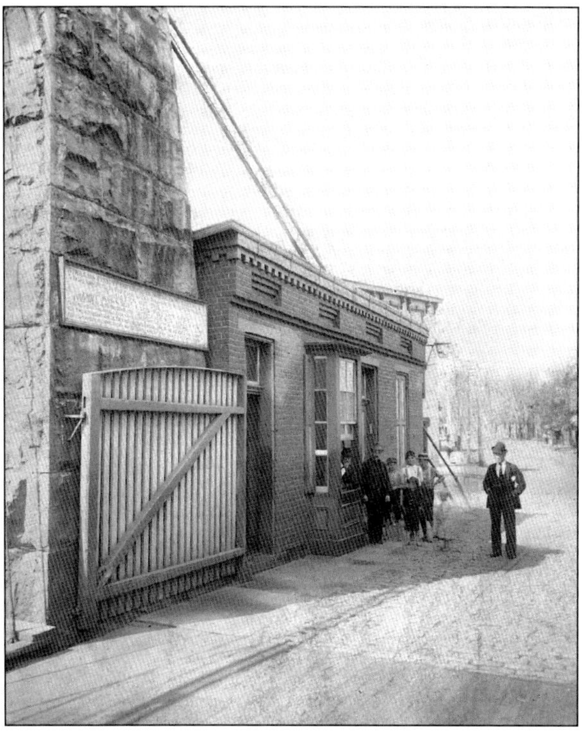

Examples of steam locomotives manufactured by Smith & Dawson in New Haven are featured at the bottom of the circular above, distributed in the 1870s. The firm, also known as the National Locomotive Works, employed nearly 200 workers and was located on First Street near the suspension bridge, which is also seen here. The freight cars, seen to the right of the bridge, belonged to the Pittsburgh & Connellsville Railroad.

The first bridge across the Youghiogheny River at Connellsville was built by Isaac Meason and Zachariah Connell in 1814. There were two subsequent spans, both of which were either destroyed by floods or collapsed due to overweight vehicles. Seen here is the tollbooth for the fourth bridge to cross the river, which was erected in 1862 in New Haven.

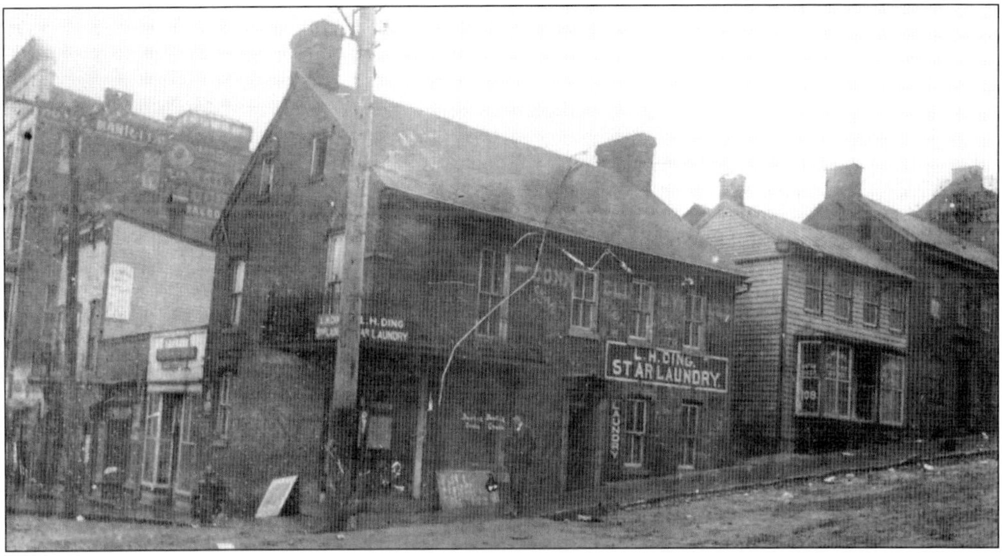

Fire and brimstone fell on this corner after Thomas Ewing established a store here in 1845. The corner reportedly received its name from heated political discussions that took place in the establishment. Brimstone Corner was subsequently occupied by the Frisbee store and remained the most desirable business site in the city for years. This photograph was taken shortly before the property was acquired by the Title & Trust Company in 1899. (Courtesy of Felix Prestia.)

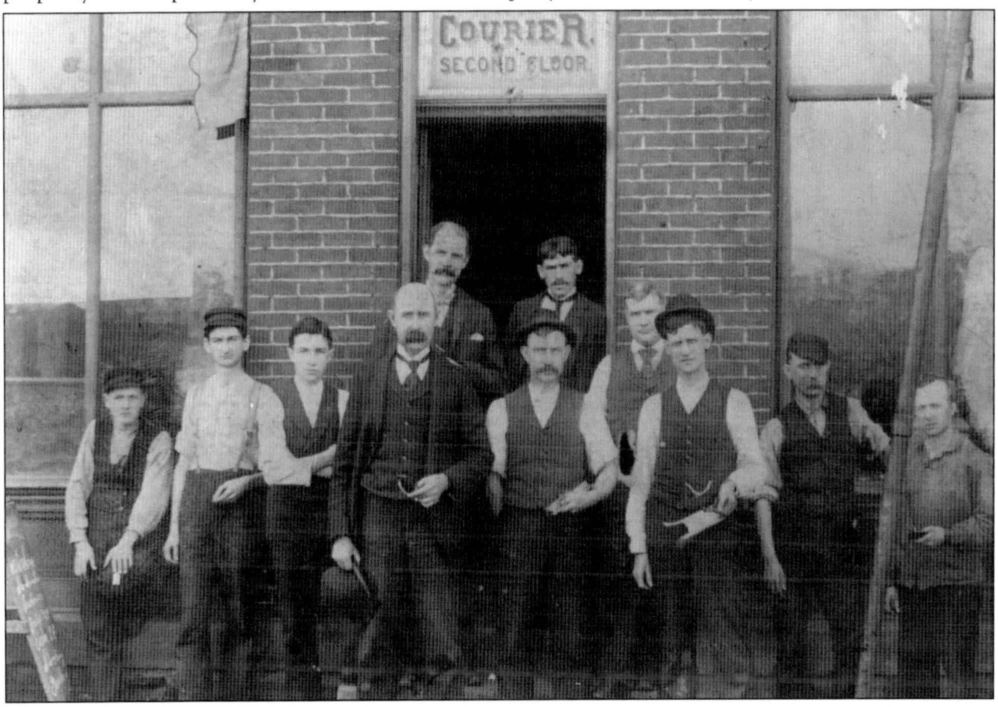

Employees of the *Daily Courier* stand next to their office on Water Street in 1895. The editor, Henry Snyder, is leaning against the doorway on the left. Snyder started writing a weekly column on the coke industry, which is generally credited with introducing the nation to the term "Connellsville Coke Region," a designation used today in books and articles relating to a unique chapter in the economic history of the United States.

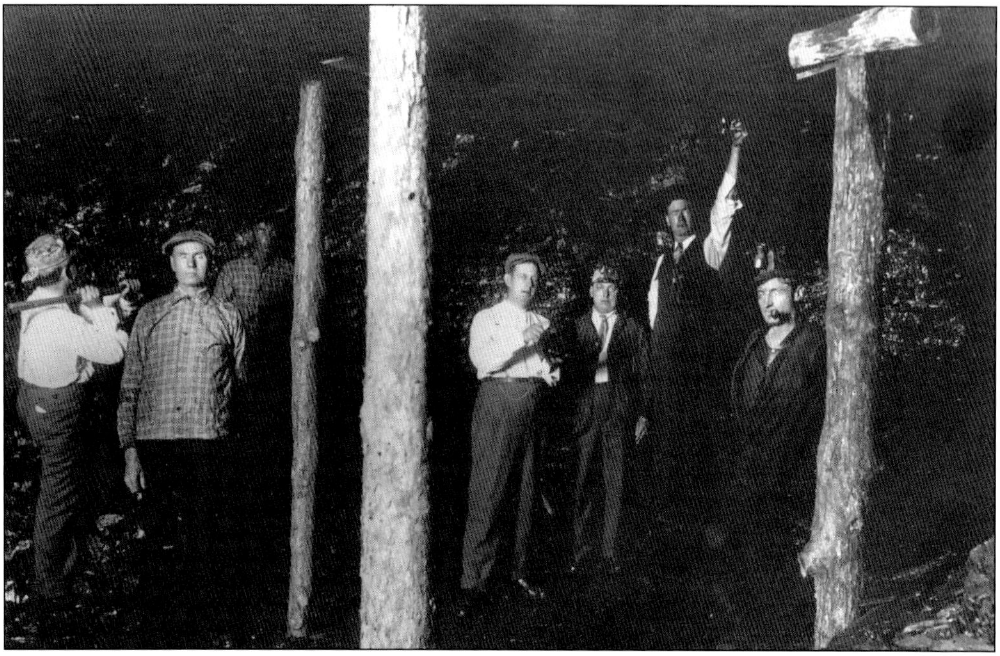

Mine bosses test for methane gas in the Pittsburgh coal seam, which averaged nine feet in thickness beneath the surface of Fayette and Westmoreland Counties. In 1890, geologists called this coal bed the most valuable mineral deposit in the world.

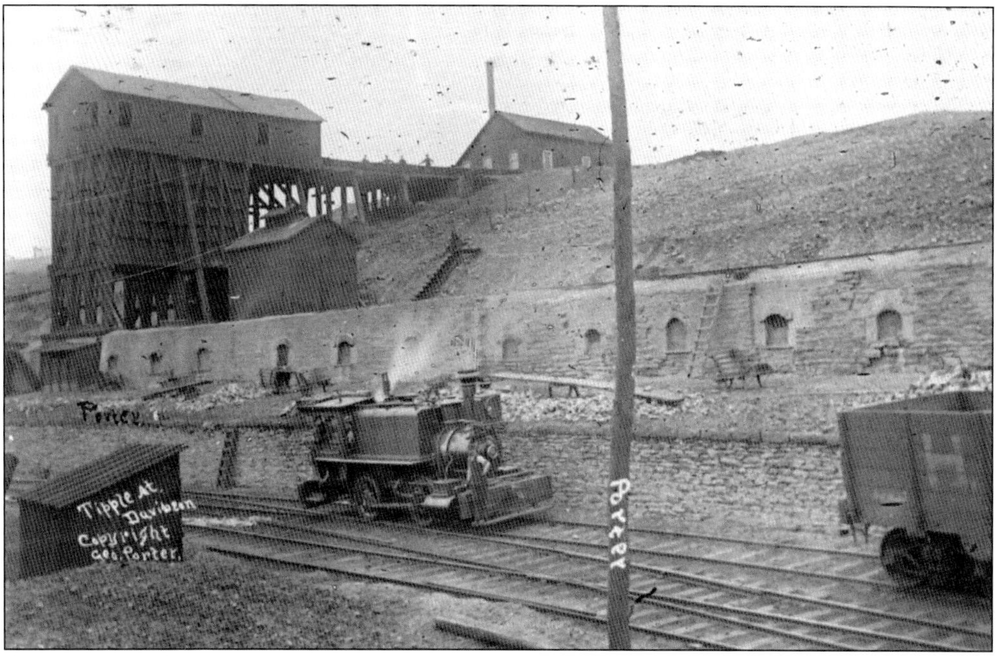

Seen here are the tipple and coke ovens at the Davidson Mine, located northwest of the city in Connellsville Township. It was established in 1856 and purchased by the H.C. Frick Coke Company in 1888. During the 1894 national strike, Frick's chief engineer, James Paddock, took refuge in the tipple after he was beaten by strikers. His body was subsequently found below the tipple with a bullet in his head. Four strikers were also killed during the violence.

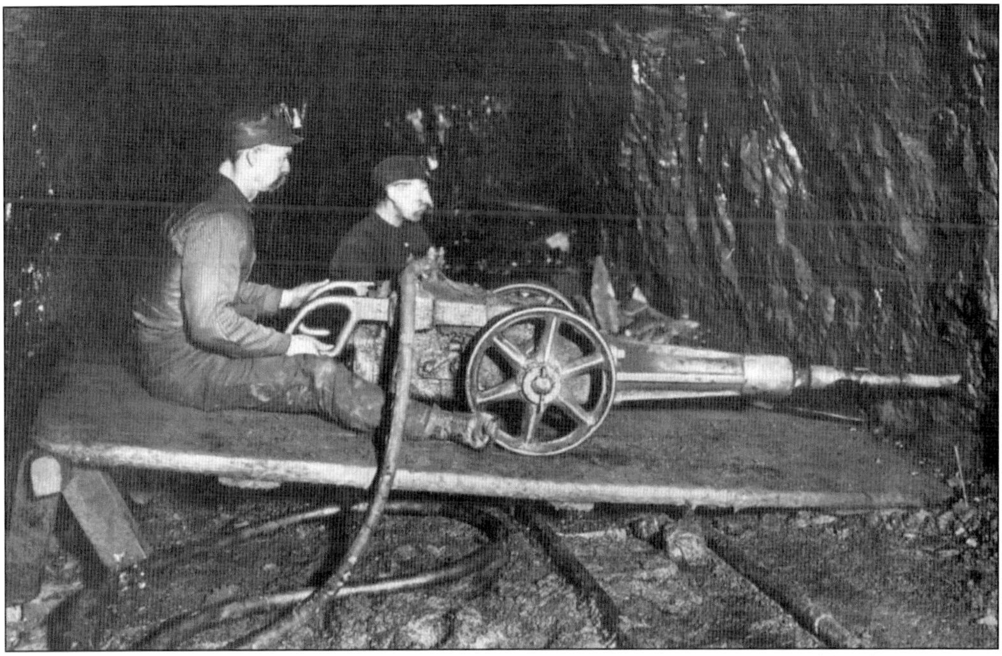

Mining was a pick-and-shovel operation from Colonial times until equipment like this Harrison undercutting machine was manufactured in the 1870s. According to Carmen DiCicio's 1996 book *Coal and Coke in Pennsylvania*, one machine operated by two workers could undercut four to six rooms of coal in a 10-hour shift. The same work required 20 pick miners. (Courtesy of Tom Korba.)

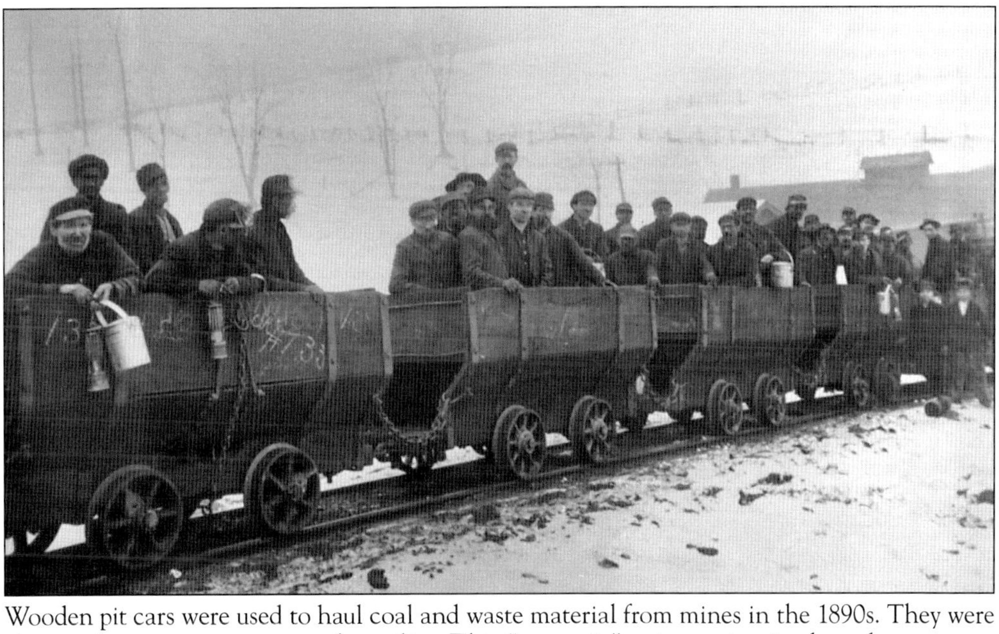

Wooden pit cars were used to haul coal and waste material from mines in the 1890s. They were also used to transport men and supplies. This "man trip" exits a mine in the coke region at the end of a shift. Note the first miner on the left, who is holding his lunch bucket and safety lamp. Drinking water was stored in the bottom of the container, while the upper sections held nonperishable food items.

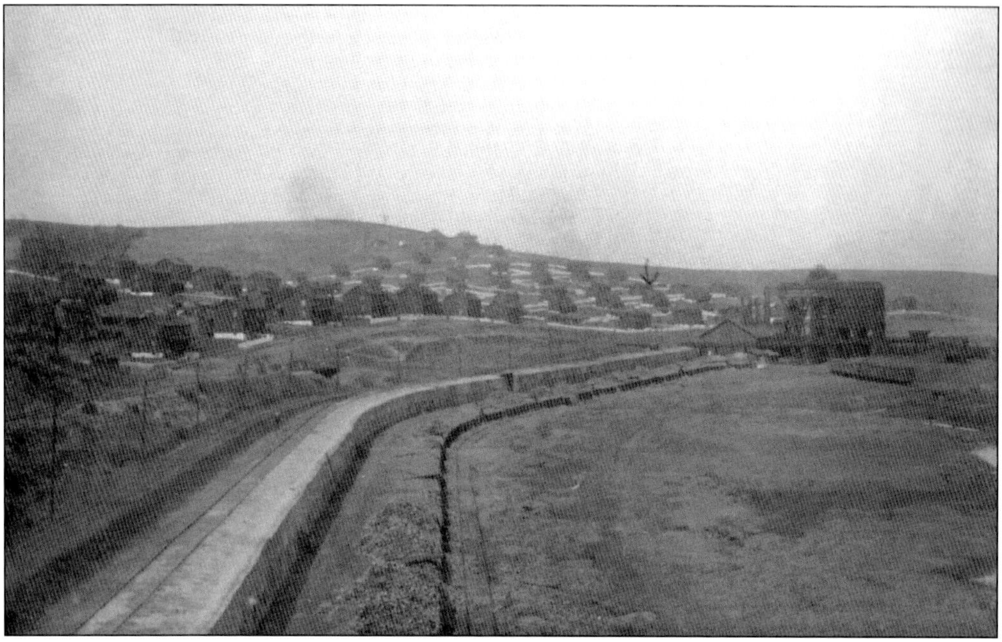

Seen here are the mine shaft, coke ovens, and company houses at Leisenring No. 1. The industrial complex was established by the Connellsville Coke & Iron Company in 1880. It was named for company president John Leisenring, a coal and transportation magnate from Philadelphia. The works contained 499 ovens. This photograph was taken shortly after the property was acquired by the H.C. Frick Coke Company in 1889. (Courtesy of Denny Gaal.)

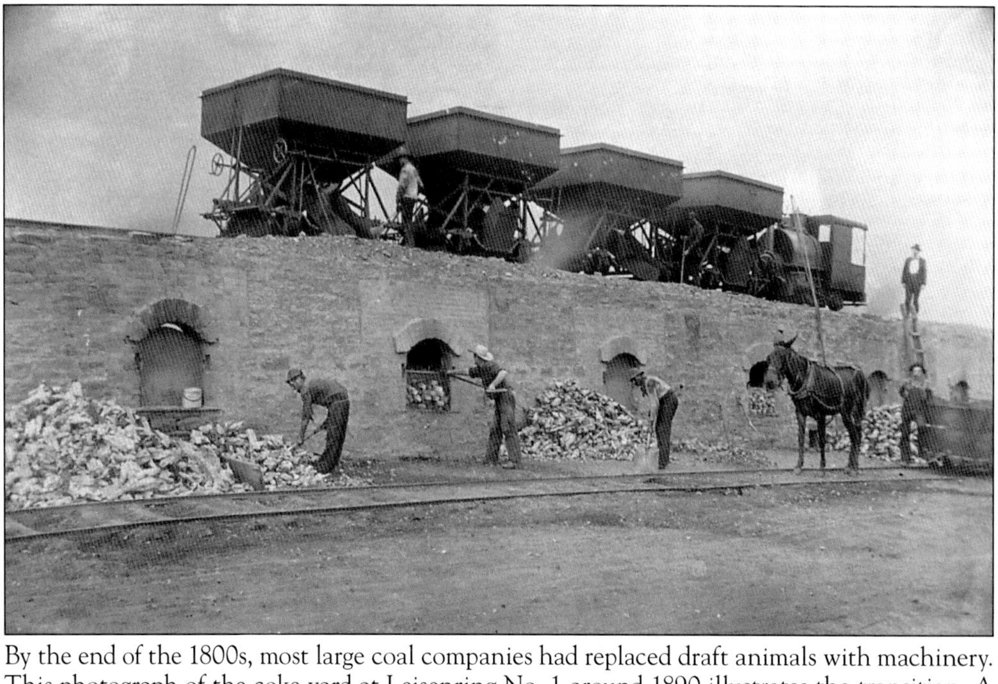

By the end of the 1800s, most large coal companies had replaced draft animals with machinery. This photograph of the coke yard at Leisenring No. 1 around 1890 illustrates the transition. A small steam engine called a dinkey is pushing four larries in position to charge the ovens, while a horse pulls wagons loaded with waste or the finished product.

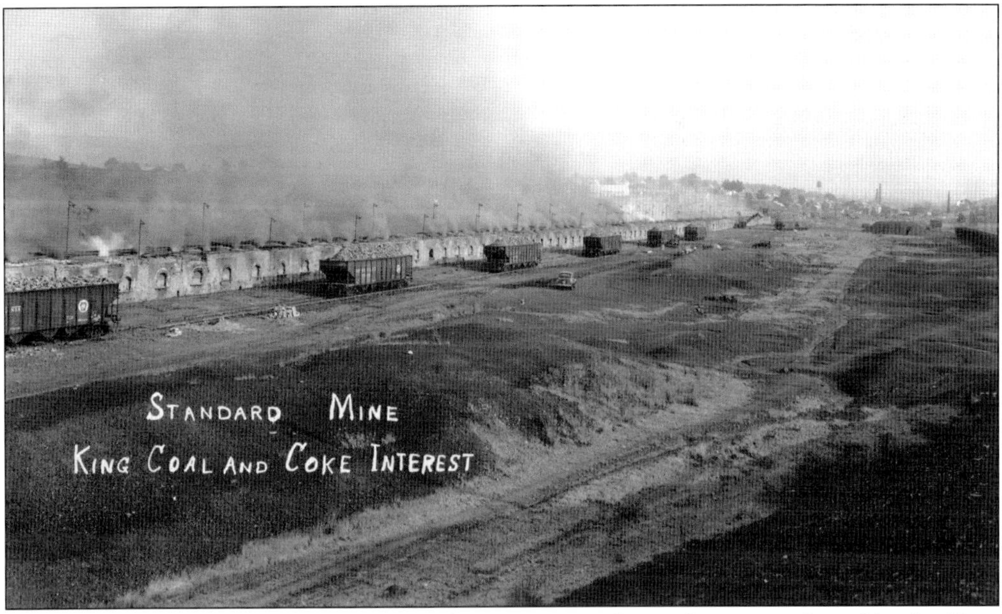

The Standard Mine, adjacent to Mount Pleasant, is seen here when it was leased by the King Coal & Coke Company from the H.C. Frick Coke Company in the 1930s. The coke yard was the largest in the world, with 999 ovens. Standard was also the largest shaft mine in the region. It extended underground to Hecla, a distance of four miles.

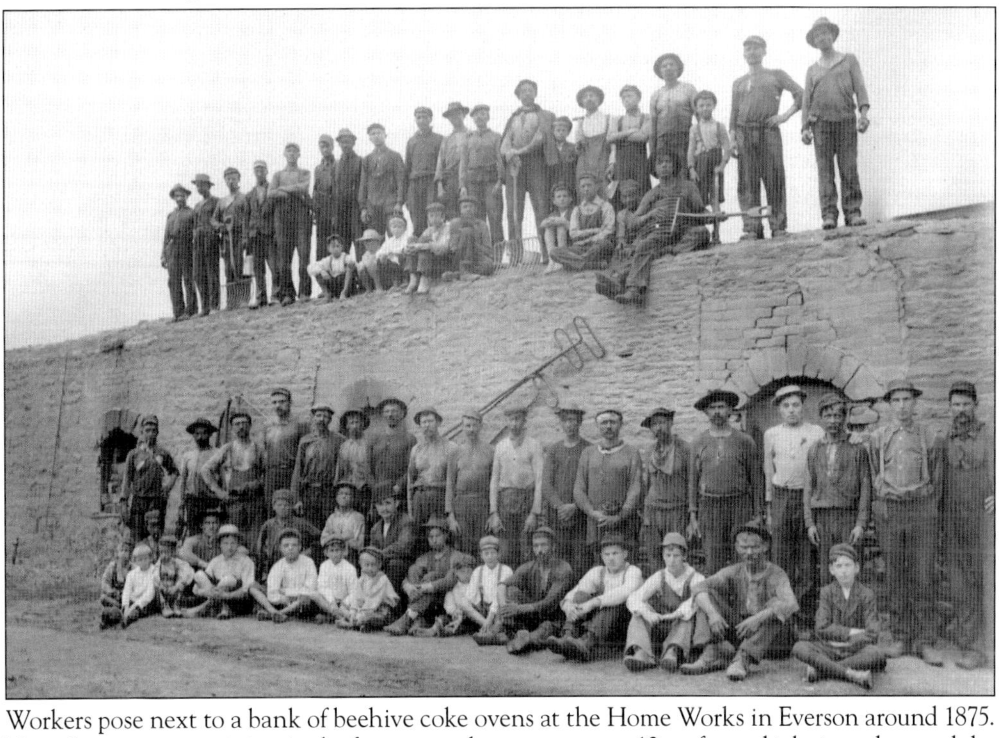

Workers pose next to a bank of beehive coke ovens at the Home Works in Everson around 1875. Note the young men sitting in the front row—boys as young as 10 performed jobs in and around the mine. They usually started working as their father's helper in order to increase family income. Child labor did not disappear until the early 1930s. (Courtesy of the *Scottdale Independent Observer*.)

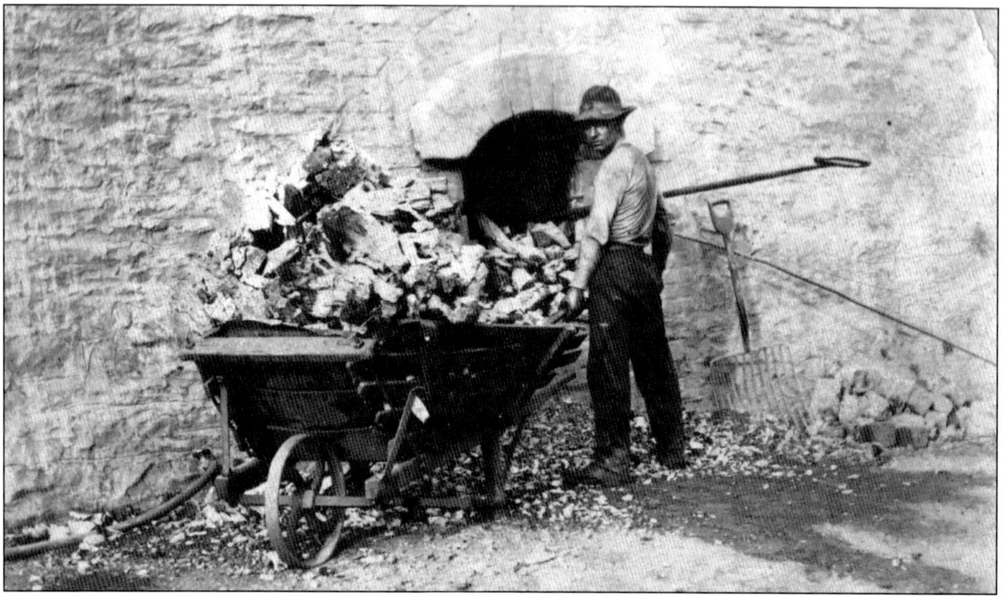

The process of drawing and loading coke from a beehive oven was a long and arduous process. Here, around 1880, a coke yard worker in the Connellsville region loads coke into a wheelbarrow. The tool, partially extended from the oven, was used to remove the coke after quenching it with water. (Courtesy of Paula Salatino Beucher.)

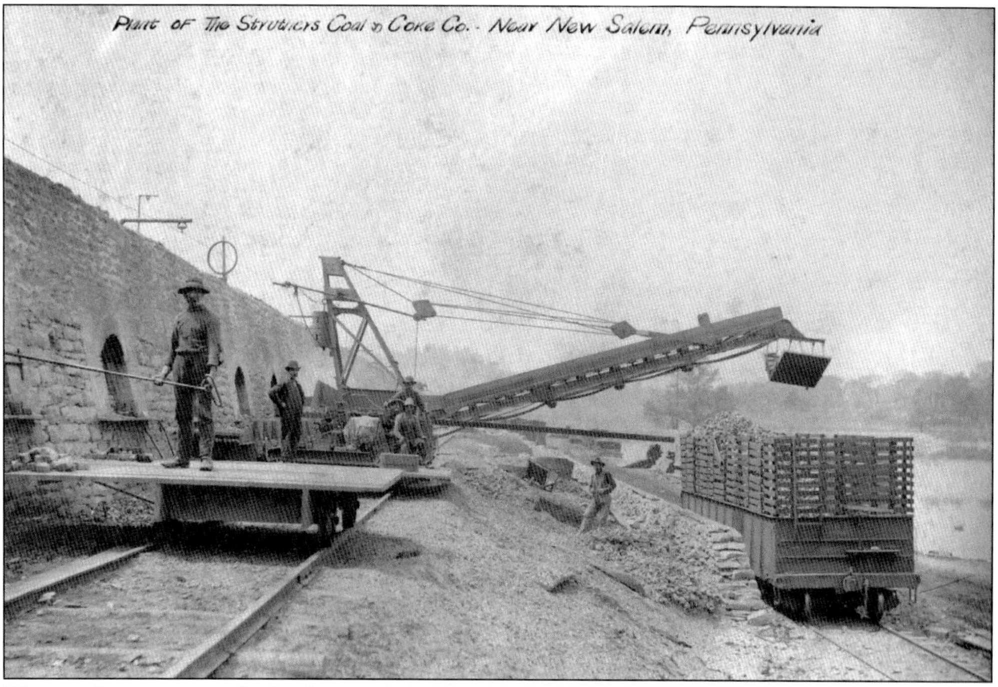

The machine seen in the background with a conveyor was used by the Struthers Coal & Coke Company around 1905 near New Salem to replace the manual drawing of coke. According to Kenneth Warren's 2001 book on the Connellsville coke industry, *Wealth, Waste, and Alienation*, five men operating the machine could draw and load 30 ovens, whereas the traditional hand methods required one man for every three ovens.

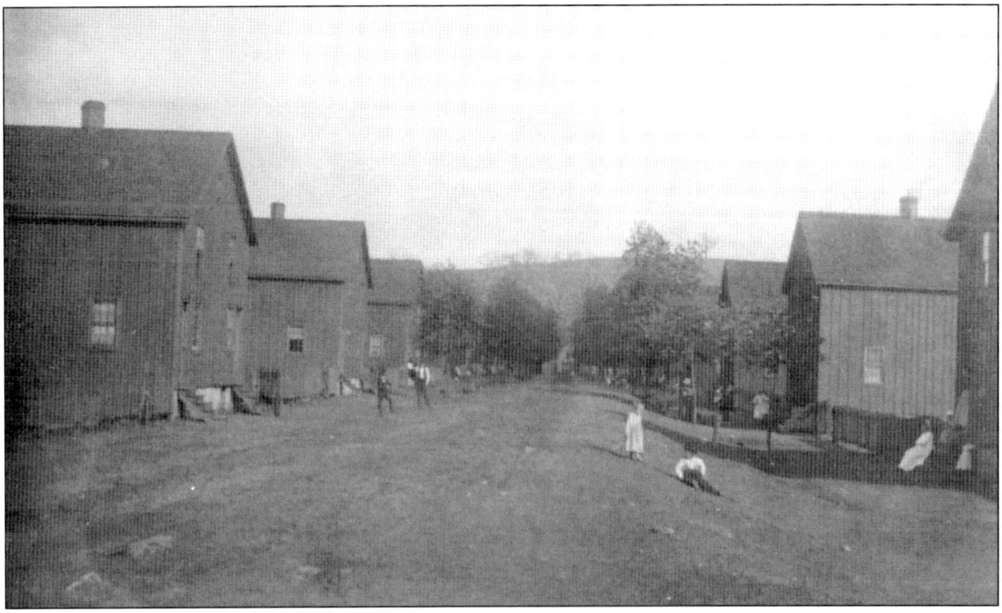

Since many coal mines were located in remote areas that lacked adequate housing, transportation, and other services, companies built towns consisting of family homes and tenements in order to attract, retain, and control their workers. Seen here is housing built in Trotter by the Connellsville Gas & Coal Company in 1880. The mine and the town were purchased by the H.C. Frick Coke Company in 1883. (Courtesy of Denny Gaal.)

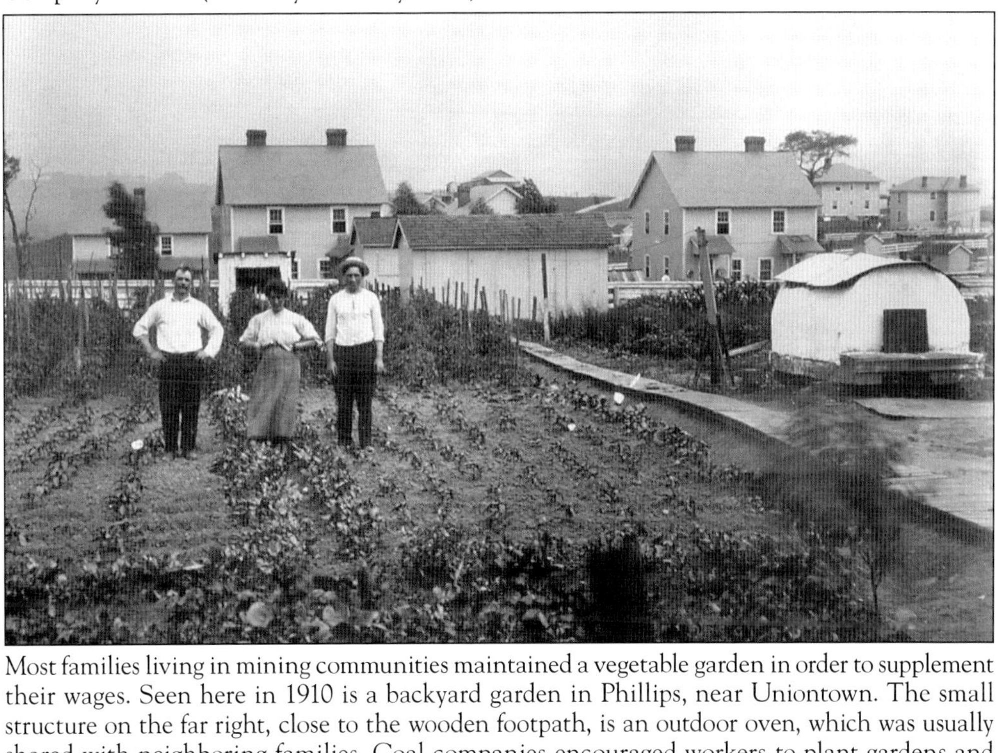

Most families living in mining communities maintained a vegetable garden in order to supplement their wages. Seen here in 1910 is a backyard garden in Phillips, near Uniontown. The small structure on the far right, close to the wooden footpath, is an outdoor oven, which was usually shared with neighboring families. Coal companies encouraged workers to plant gardens and awarded prizes for the best vegetables and flowers.

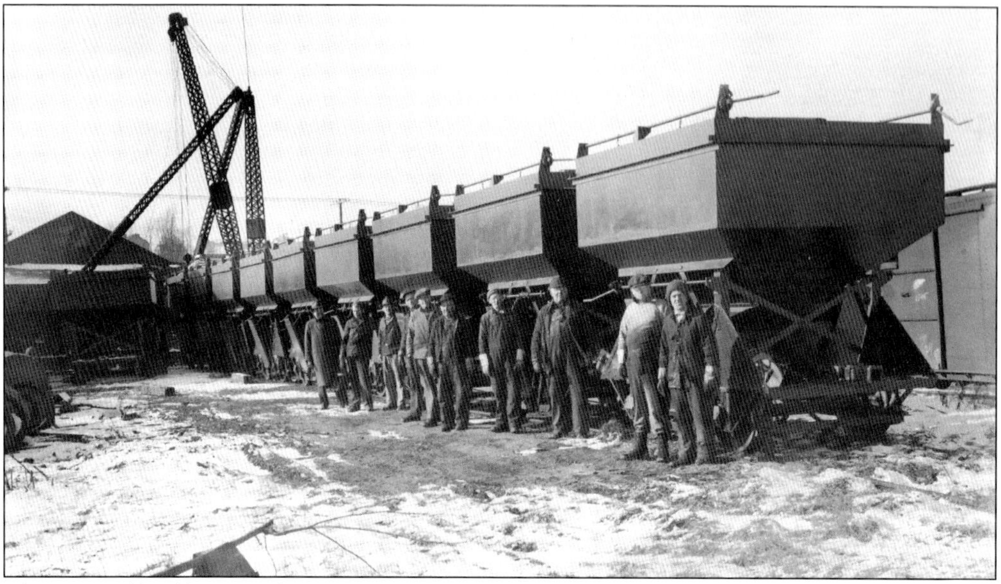

The Connellsville Manufacturing & Mine Supply Company produced these larries, which were used to charge coke ovens. In addition to this product, the company made cages, hoisting machines, pumps, fans, and compressors for the mining industry. It was organized in 1901 under the leadership of president Rockwell Marietta, vice president Clair Stillwagon, and secretary-treasurer W.F. Soisson. The company was on South Fourth Street in New Haven.

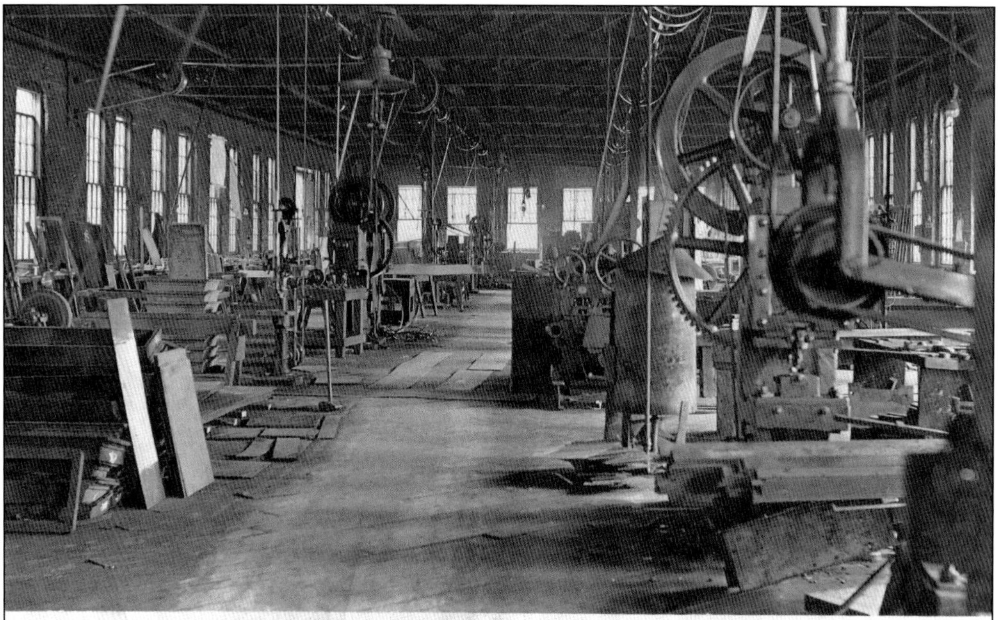

The interior of the Pittsburgh Safe Company is seen here after it was organized in 1902. The company was located in South Connellsville and employed close to 100 residents. In addition to this industry, South Connellsville was the location of the Soisson brickworks, the Humbert Tin Plate Company, and other factories that manufactured locks, steam-powered automobiles, glassware, and metal castings.

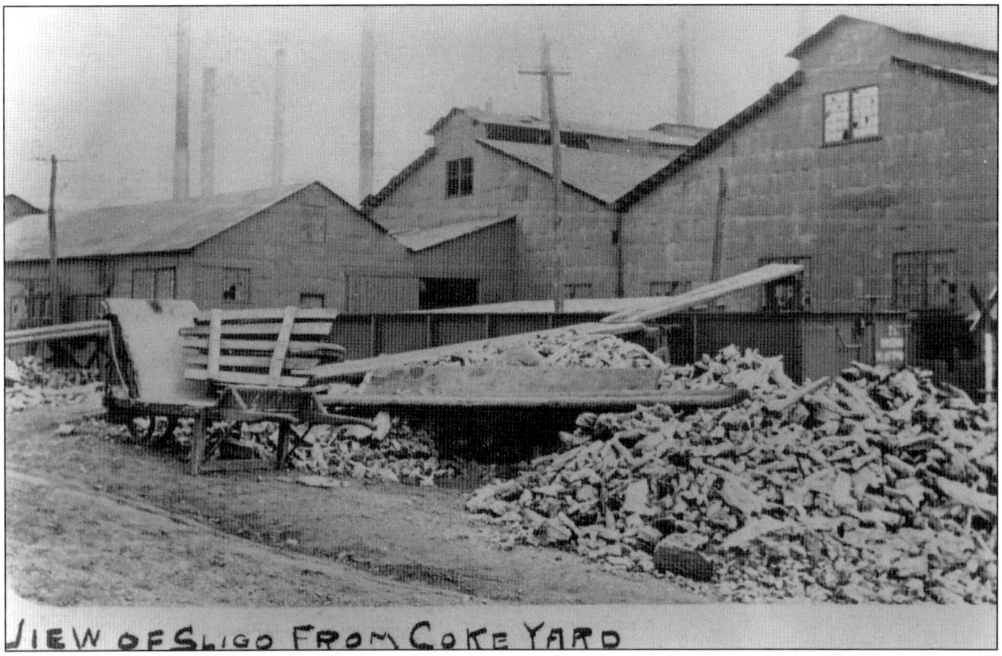

The name of a baseball field in the north end is all that remains of the Sligo Iron & Steel Company, which moved from Pittsburgh to Connellsville in 1902. It employed between 300 and 400 workers who manufactured iron plates, bars, and rails. Neighboring industries included the Munson Heater Company and the Davidson Mine.

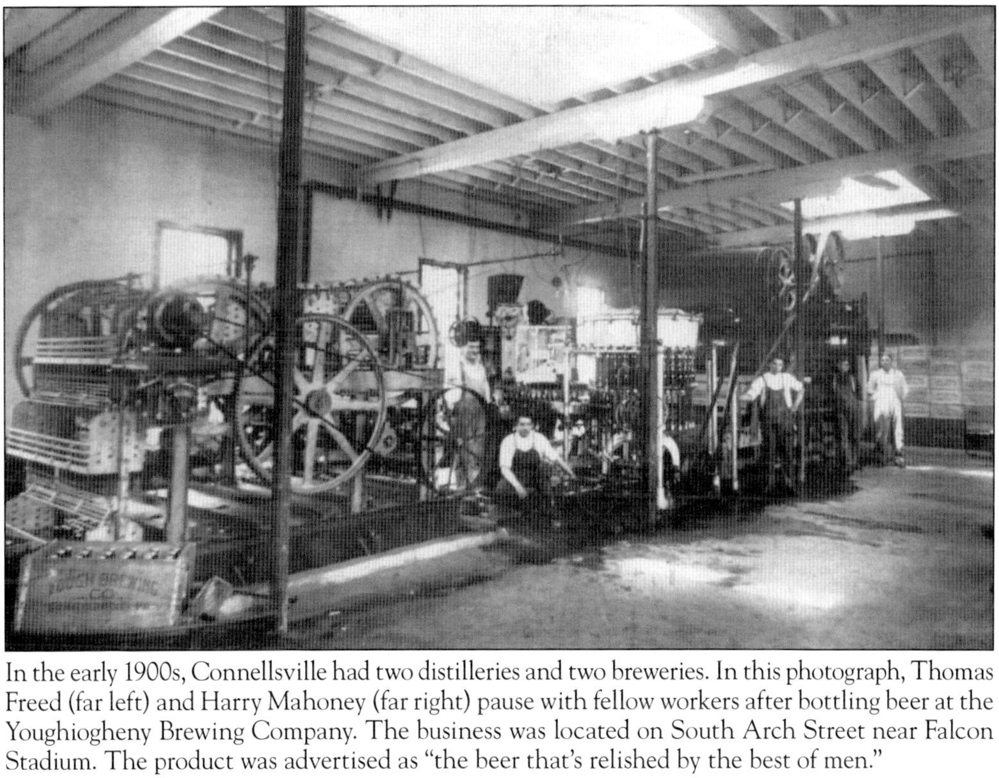
In the early 1900s, Connellsville had two distilleries and two breweries. In this photograph, Thomas Freed (far left) and Harry Mahoney (far right) pause with fellow workers after bottling beer at the Youghiogheny Brewing Company. The business was located on South Arch Street near Falcon Stadium. The product was advertised as "the beer that's relished by the best of men."

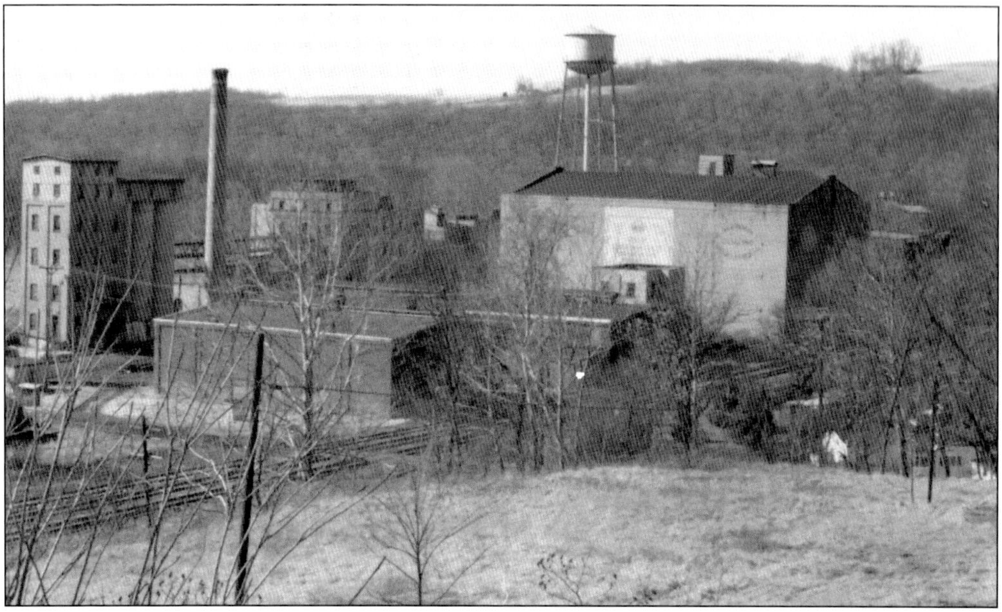

During Colonial times, most farmers operated small distilleries, but Abraham Overholt of West Overton established a commercial enterprise to market his rye whiskey. In 1854, his son Henry started a second distillery at Broad Ford, which remained open during Prohibition with a license to produce Old Overholt as a medicinal spirit. The modernized Overholt Distillery is seen here during the time it was operated by National Distillers, from 1930 to 1987.

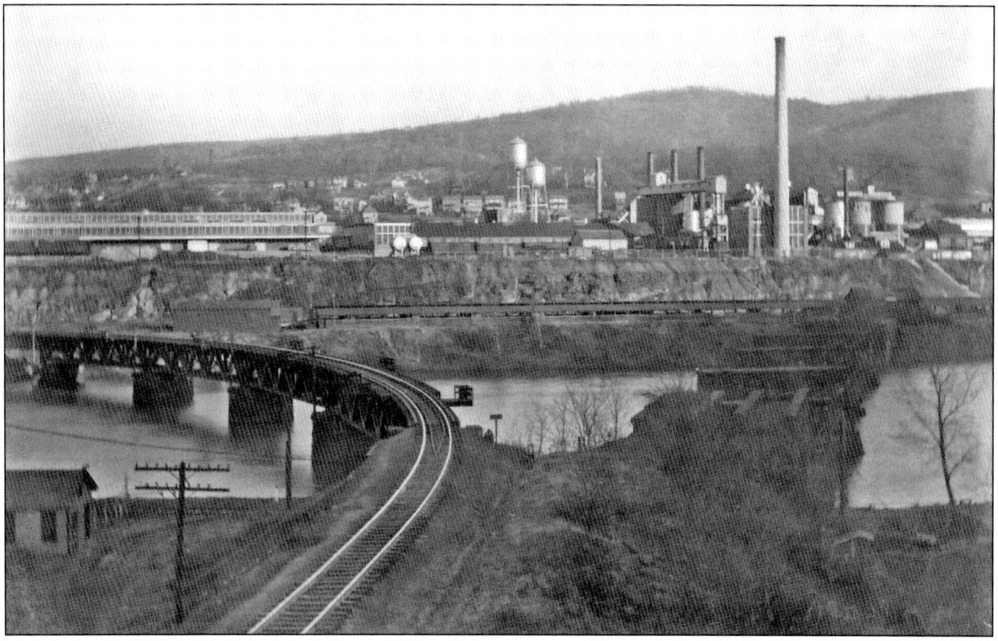

Anchor Hocking Glass Corporation is seen here across the river and below the Chestnut Ridge. Reportedly, most people born in Connellsville worked part of their lives at Anchor. After World War II, it was the largest employer in Fayette County, with an estimated 2,400 workers. The railroad bridge (left) connected the B&O main line in Connellsville to Morgantown and Fairmont, West Virginia. It was known as the Sheepskin Line.

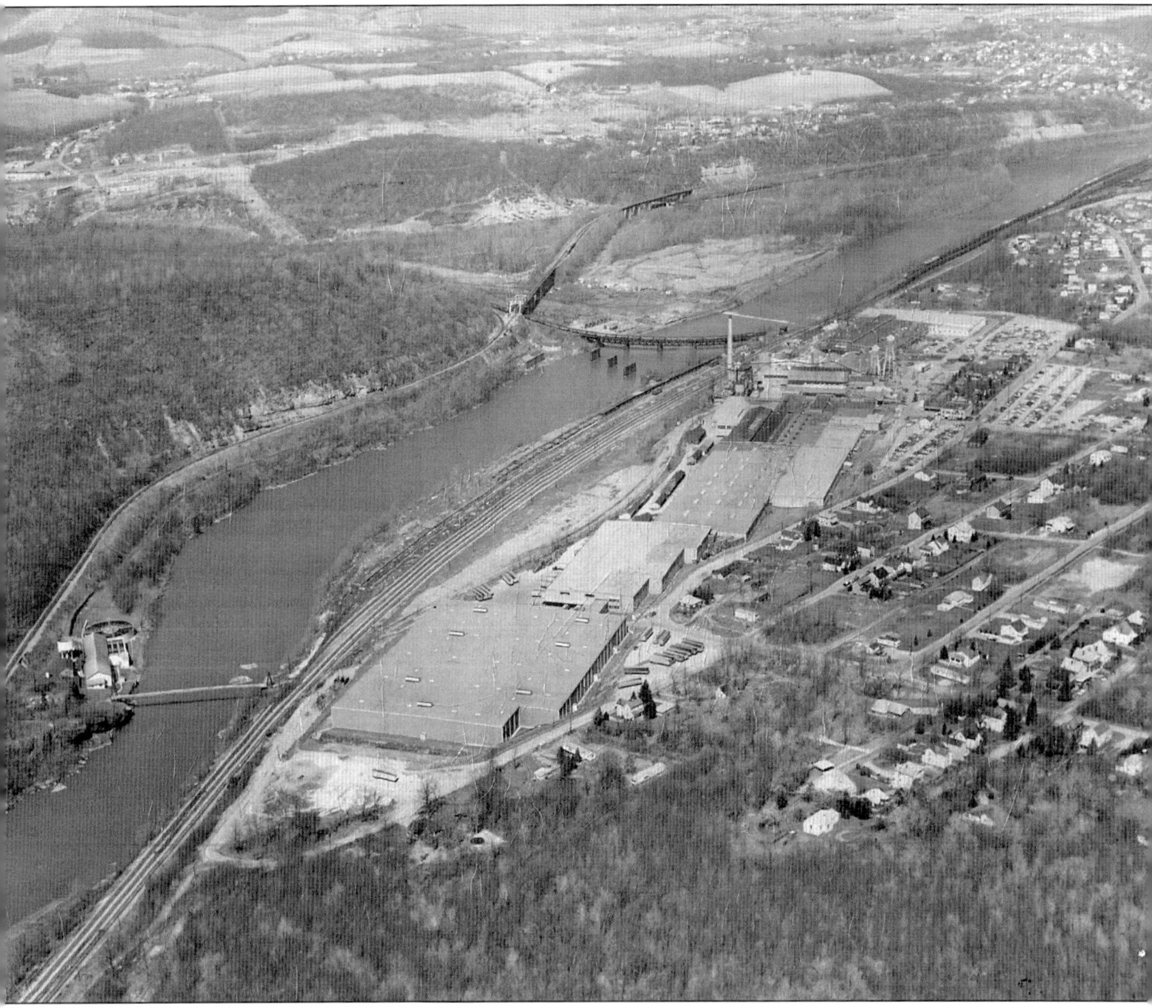

This is a 1960s view of the Youghiogheny River as it flows north from South Connellsville. The factory and warehouses of Anchor Hocking dominate the center of the photograph. The B&O Railroad is between the industrial complex and the river. The main line of the Western Maryland Railway is also visible on the left side of the river. South Connellsville was an early glass-manufacturing center. The first company to locate in the borough was Ripley Glass, established around 1911. In 1917, the property was acquired by the Anchor Cap & Closure Company (Capstan), which merged with the Hocking Glass Corporation in 1937. (Courtesy of Bill Sechler.)

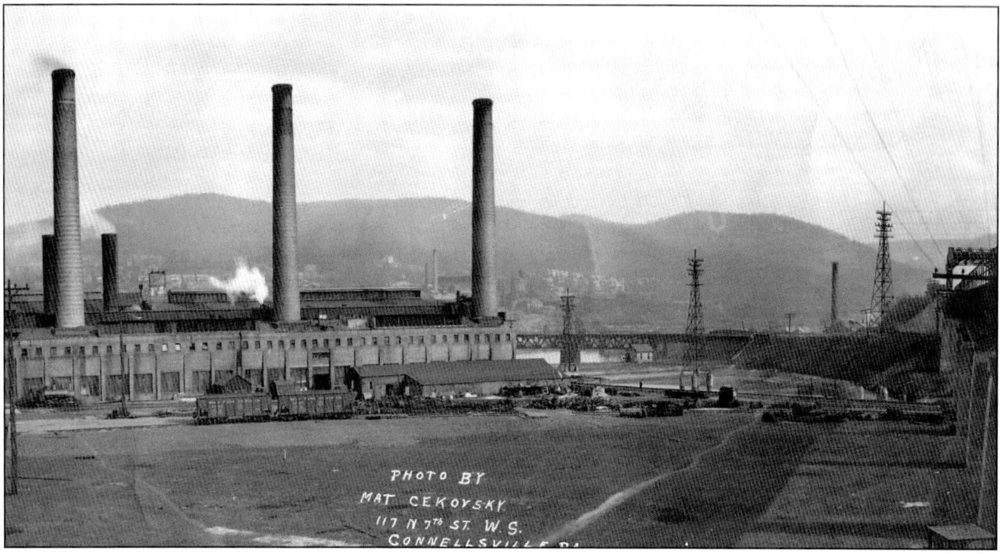

The West Penn Electric Company built this power plant on the west side of the Youghiogheny River shortly after the utility was incorporated in 1904. It supplied power to Allegheny, Fayette, Washington, and Westmoreland Counties as well as the West Penn Railways, an auxiliary company. The facility was replaced with a substation in the 1970s. Today, West Penn is a subsidiary of First Energy, which supplies electricity to customers from Indiana to the New Jersey shore.

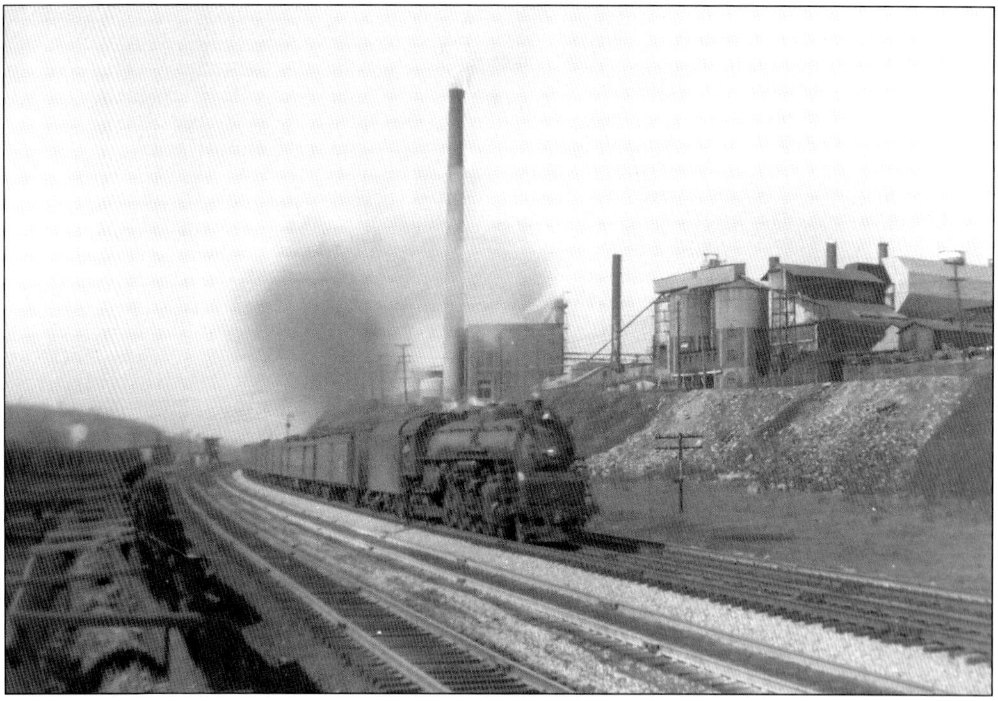

Seen here around 1959 is a B&O passenger train heading east toward Cumberland, Maryland. In the lead is a Mountain, or T3-B–class, engine, which was usually assigned to passenger service because of its speed and power. Anchor Hocking is on the right and a portion of the B&O stockyards is visible on the left. The stockyards operated from 1910 to 1967.

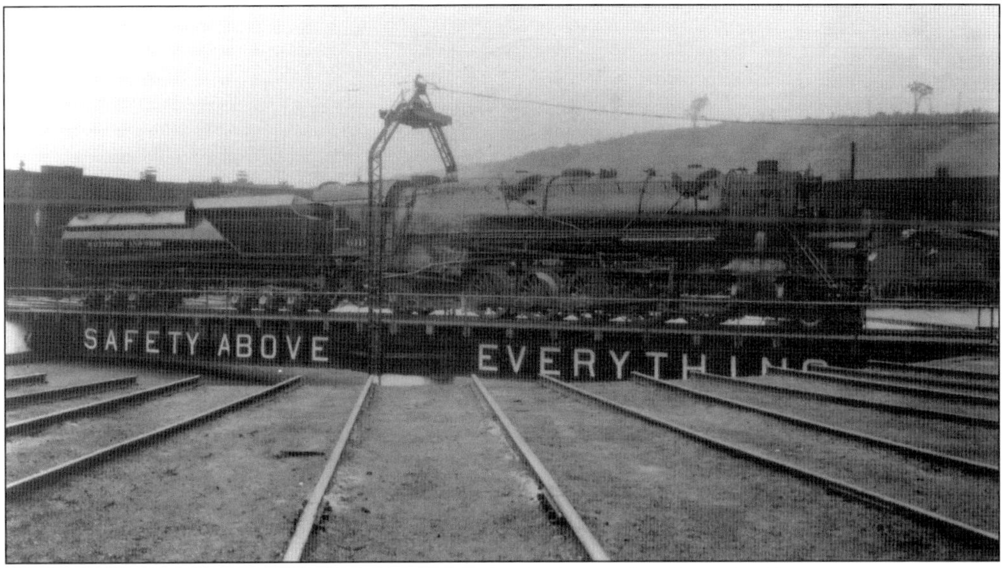

A Big Six is seen here in 1927 on the turntable in the Connellsville yard. These locomotives were used by the B&O to conquer the steep grades over the Allegheny Mountains. The roundhouse, constructed in 1903 with 24 stalls, is in the background. A fire destroyed all but five of the stalls in 1928. Although the railroad did not rebuild the original structure, maintenance work on locomotives continued in Connellsville until the late 1970s.

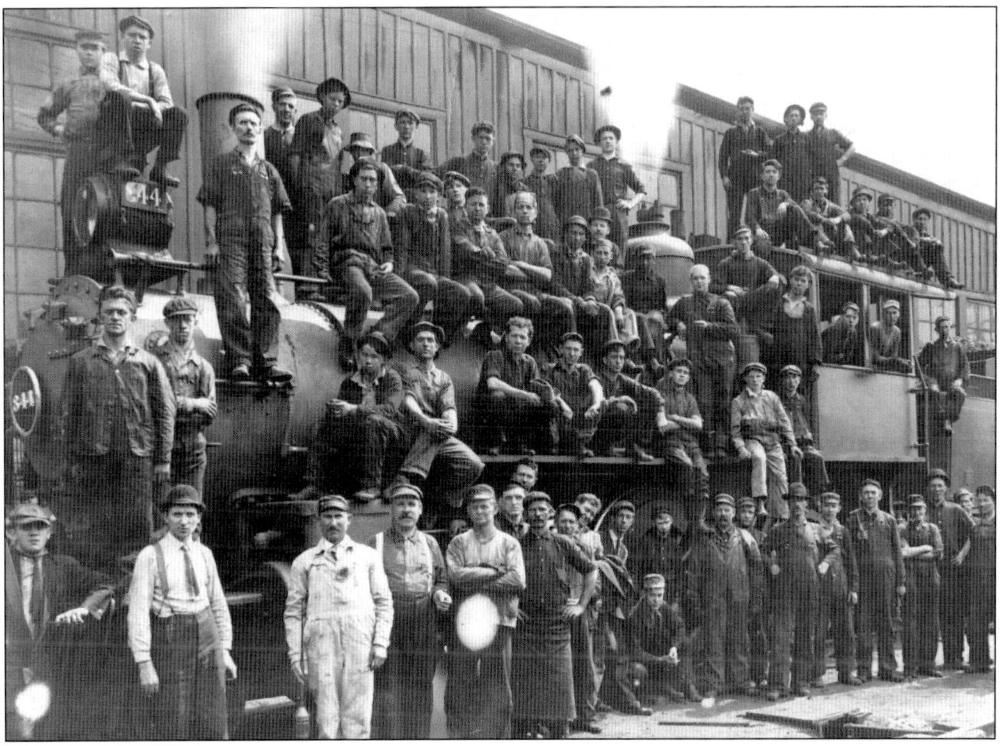

Seen here are some of the 250 men who were employed to maintain and repair steam locomotives for the B&O Railroad in Connellsville. The photograph was taken next to the roundhouse in 1911 on engine No. 844, a B&O class I-5 (American) manufactured by Baldwin Locomotive Works.

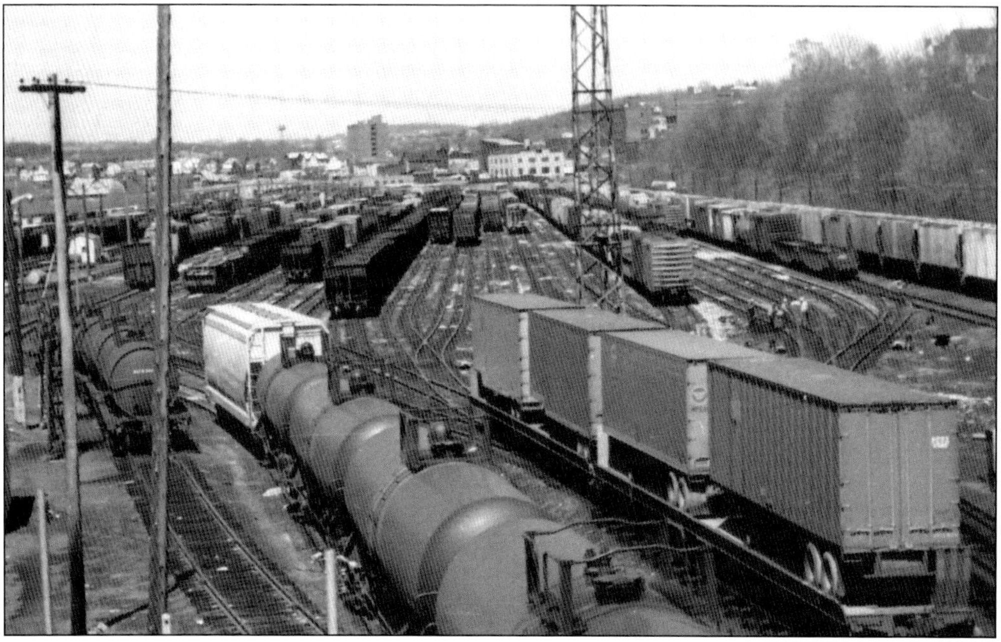

This photograph shows the B&O classification yard around 1970, looking north toward downtown Connellsville. The yard had 22 tracks and could hold 2,500 cars. During this time, there were approximately 900 employees working in the yard and on engine crews. The CSX Corporation started downsizing the yard in the mid-1980s, gradually transferring operations to other locations.

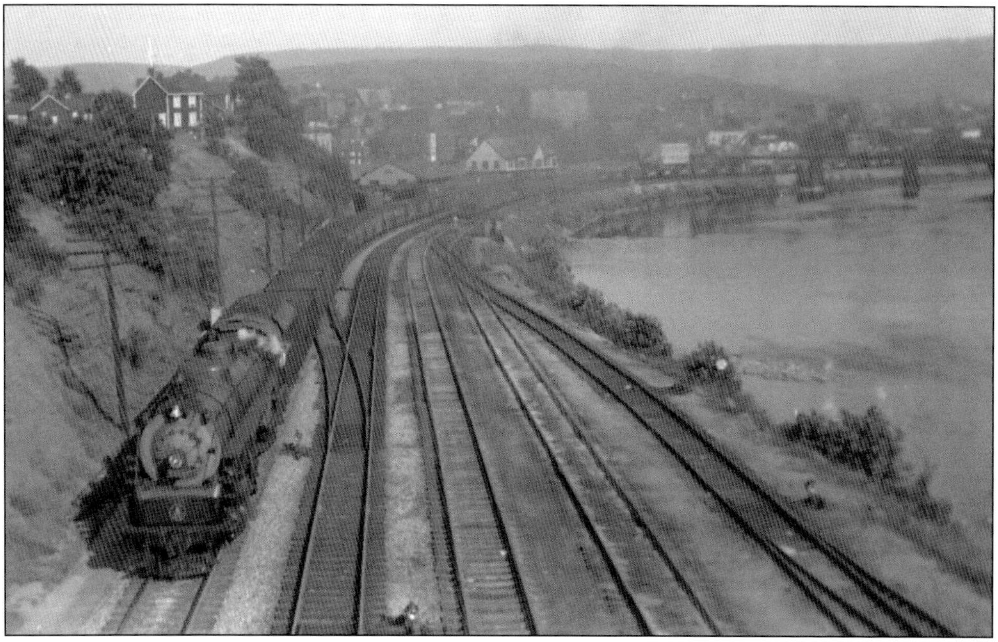

A westbound coal drag passes the B&O freight and passenger stations, which are visible behind the hoppers near the center of the photograph. The three tracks to the right were initially constructed to accommodate coke trains waiting to enter the weighting facility in the Connellsville yard. After 1904, every car loaded with coke leaving the region on the B&O Railroad had to pass through the scales in Connellsville.

Two

Living, Shopping, and Moving Around

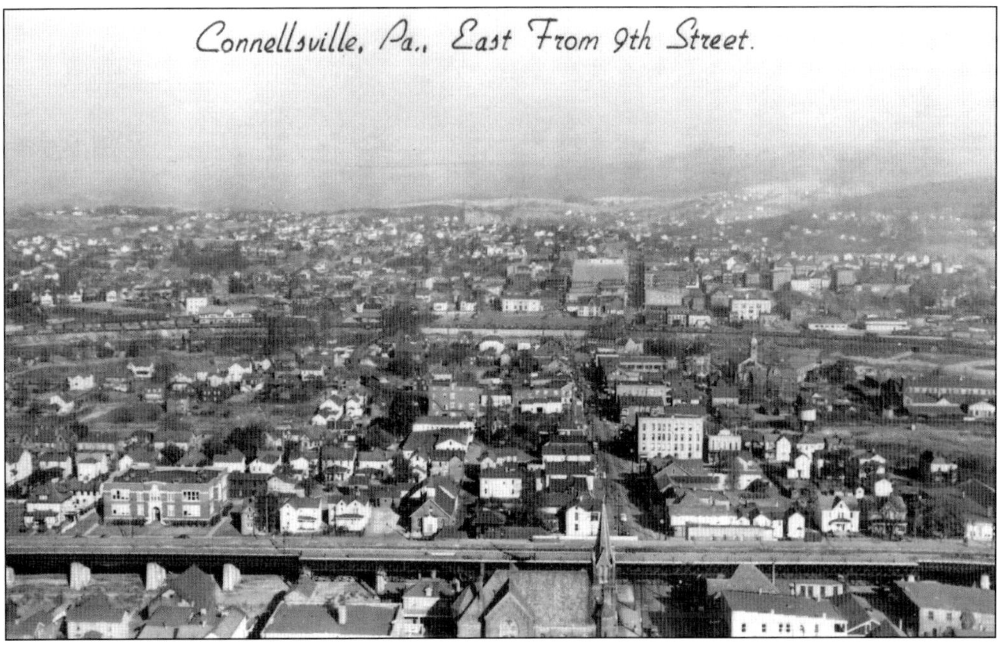

The Connellsville Chamber of Commerce issued this photographic postcard in the 1950s in order to promote the city's attractions. The back of the card read, "This is our city with 13,293 persons whose friendliness knows no bounds." In addition to many other landmarks, the card shows the Pittsburgh & Lake Erie Railroad viaduct in the foreground, directly beyond the spire of St. John the Evangelist Roman Catholic Church. (Courtesy of Margaret Molinaro.)

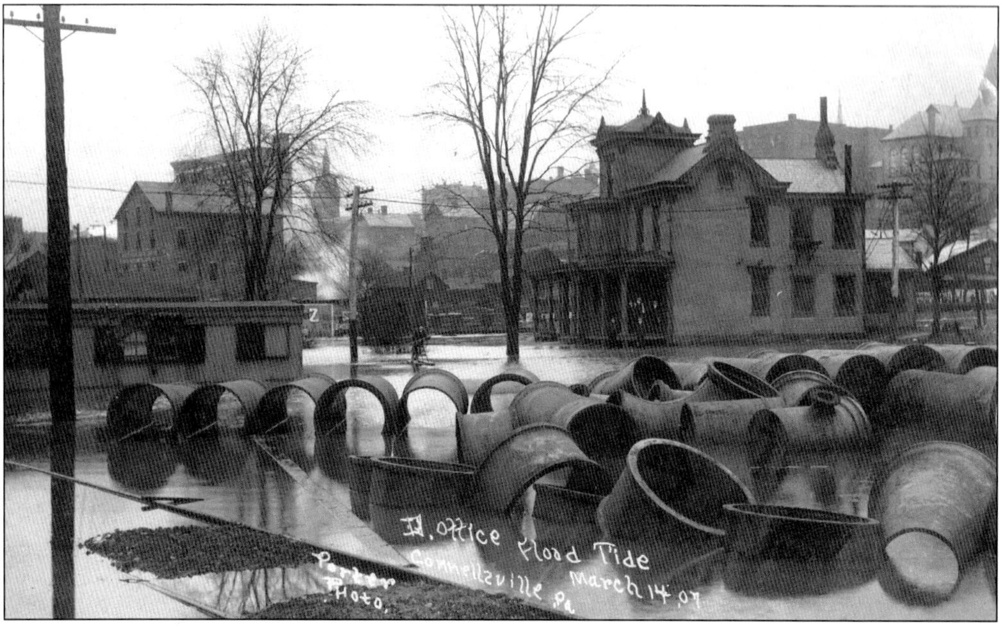

The only way to get around South Arch Street on March 14, 1907, was on improvised pontoons, as used by the man next to the tree in the center. After flooding, the Youghiogheny River current was strong enough to move these large pipes stored in the area. This scene is near the present location of Stone & Company. The house on the right is the B&O yard office. (Courtesy of Barbara and Tony Keefer.)

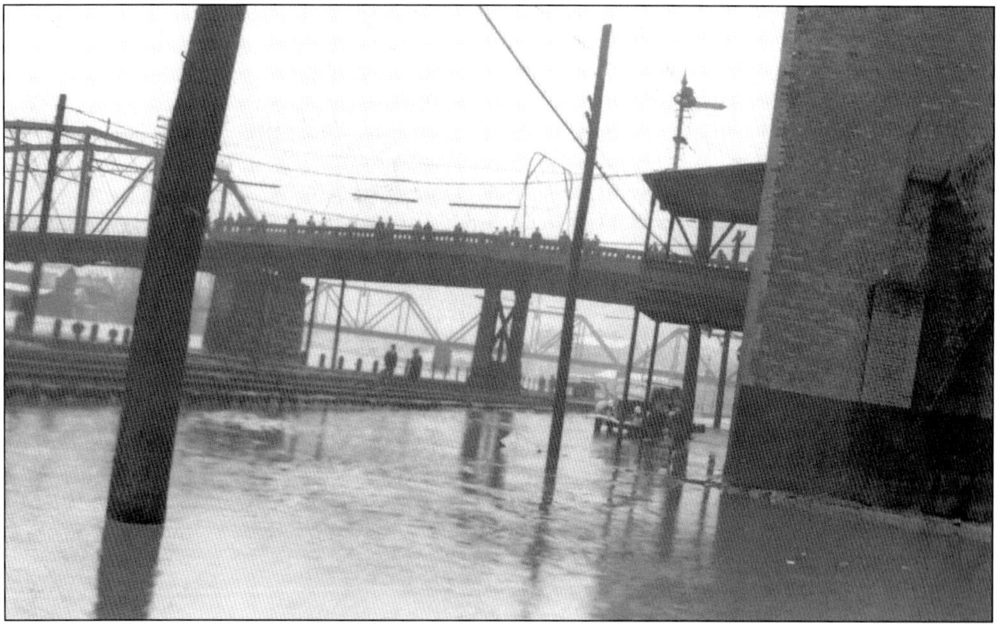

Moving around Connellsville on Water Street was difficult in 1936 after the Youghiogheny River flooded low-lying areas, including the B&O Railroad tracks. The Crawford Avenue and Pennsylvania Railroad bridges are visible in the background. Floods continued to threaten communities along the river until the Youghiogheny Dam was completed in 1943 above the town of Confluence. (Courtesy of Felix Prestia.)

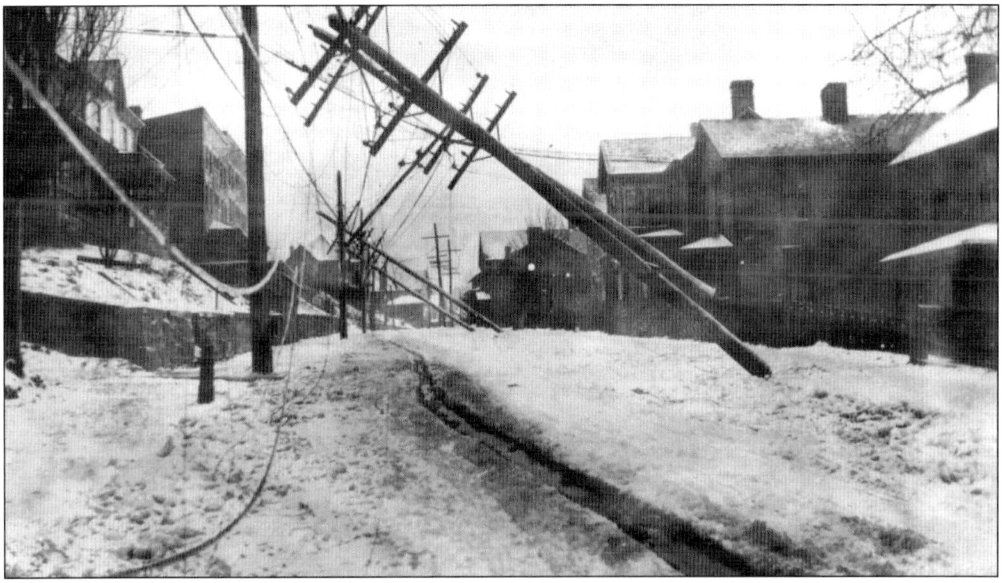

Longtime residents still remember this ice storm in February 1924. Its intensity toppled these power lines on East Fairview Avenue. Those who lived through the storm said that it was impossible to stand up on the ice or walk outside the house.

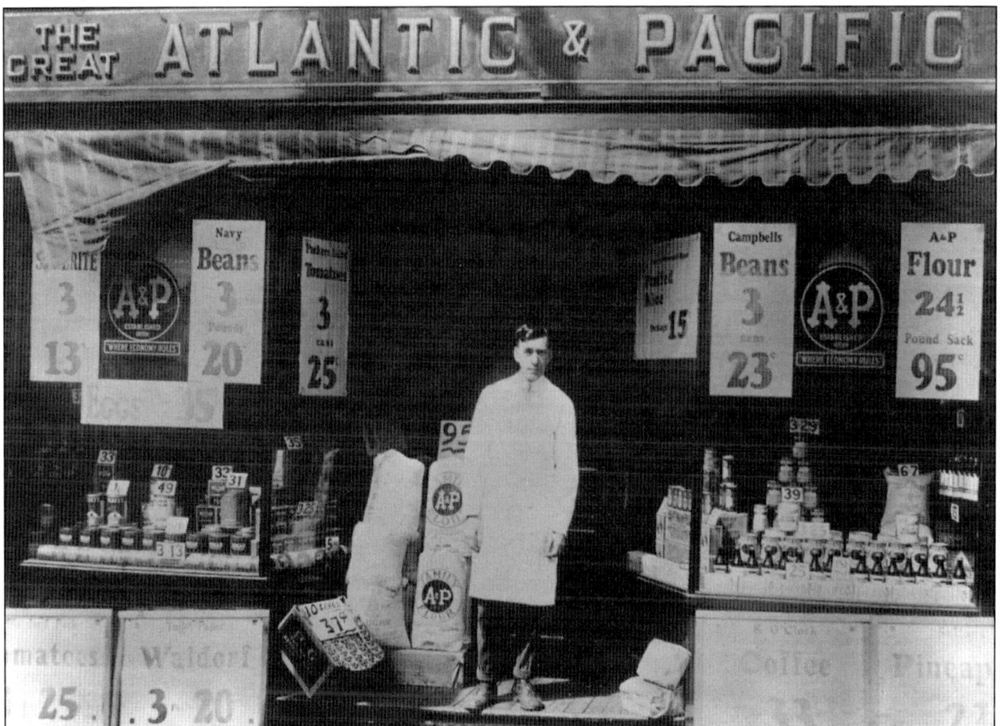

Charles Glotfelty stands in front of his Great Atlantic & Pacific Tea Co. (A&P) store in the 1930s, which was situated on what is now the parking lot at Vona's Dairy Bar on East Crawford Avenue. Glotfelty's brother Arthur managed another A&P around the same time in the north end of Connellsville. (Courtesy of Susan Rudnik.)

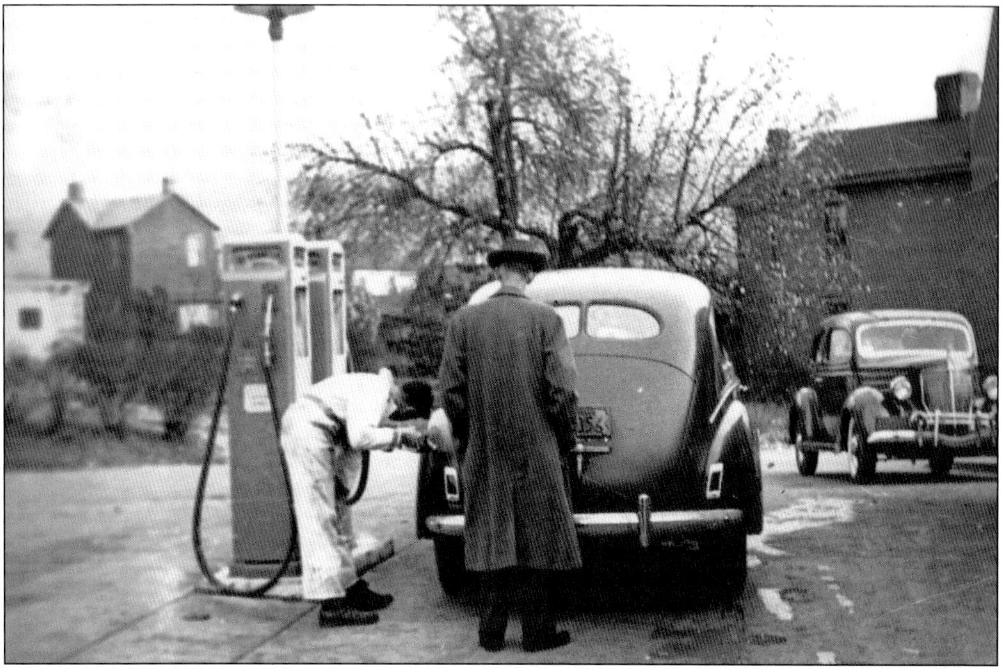

Car dealerships, auto repair shops, and gas stations were located on sections of Crawford Avenue before Route 119 was rerouted through the north end of Connellsville in 1953. This 1940s photograph shows an attendant pumping gas at Stewart's Pennzoil station on the corner of East Crawford and Cottage Avenues.

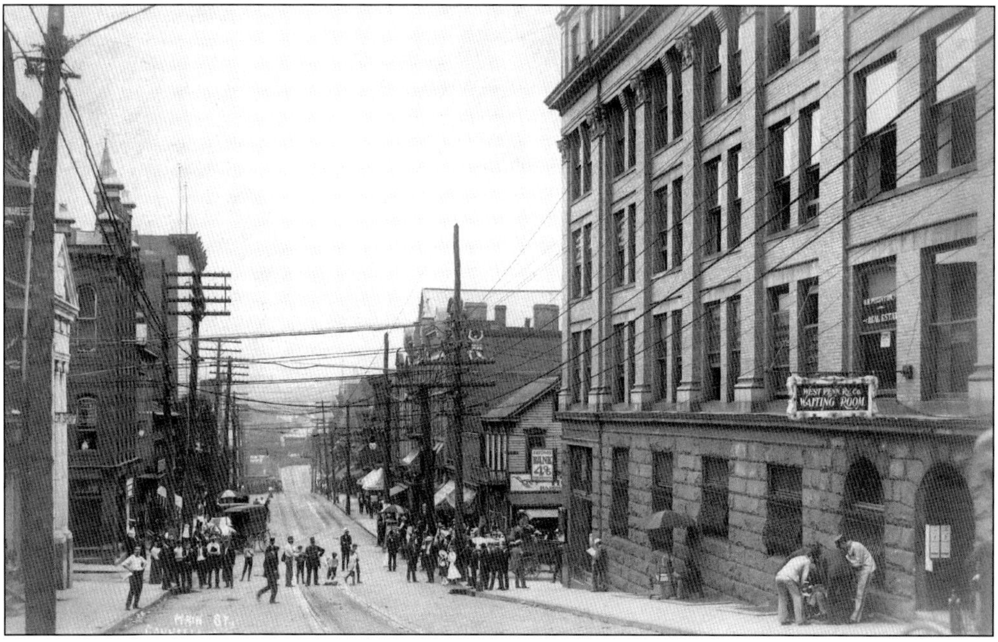

The Title & Trust Company of Western Pennsylvania (right) dominates this view of Brimstone Corner in 1904. The five-story office building, erected in 1901, was the first savings bank in the community. In addition to local businessmen such as Rockwell Marietta and John Soisson, Andrew Mellon was also a member of the board.

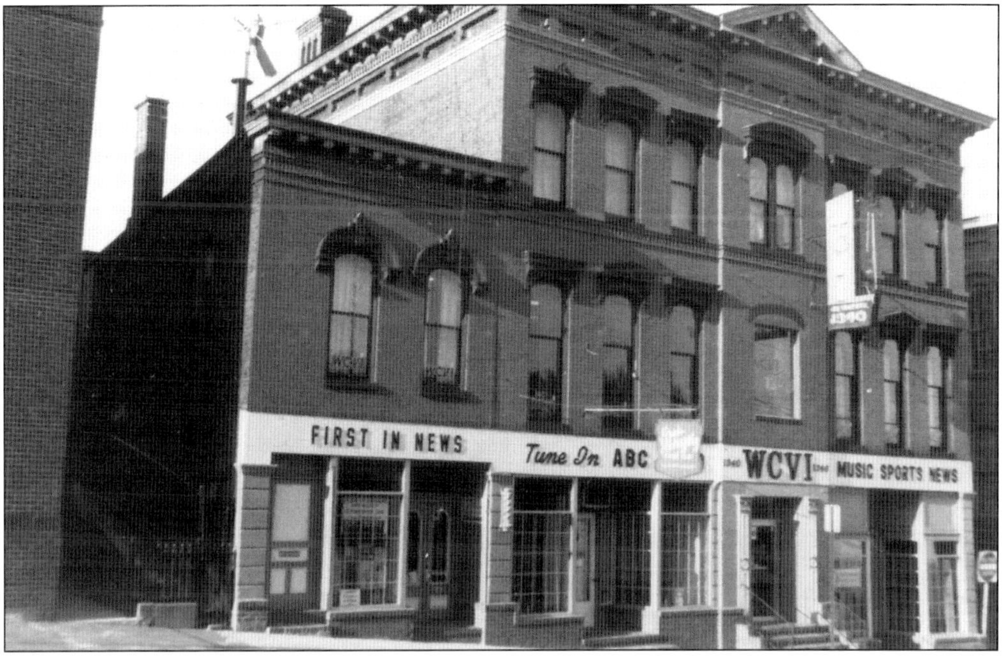

At 1340 on the AM radio dial, WCVI was the voice of Connellsville. The station operated from this studio at 133 East Crawford Avenue from 1947 to 2000. During its formative years, it was home to Dick Jensen and Donny Cable, who played the latest records and entertained groups of teenagers in the studio.

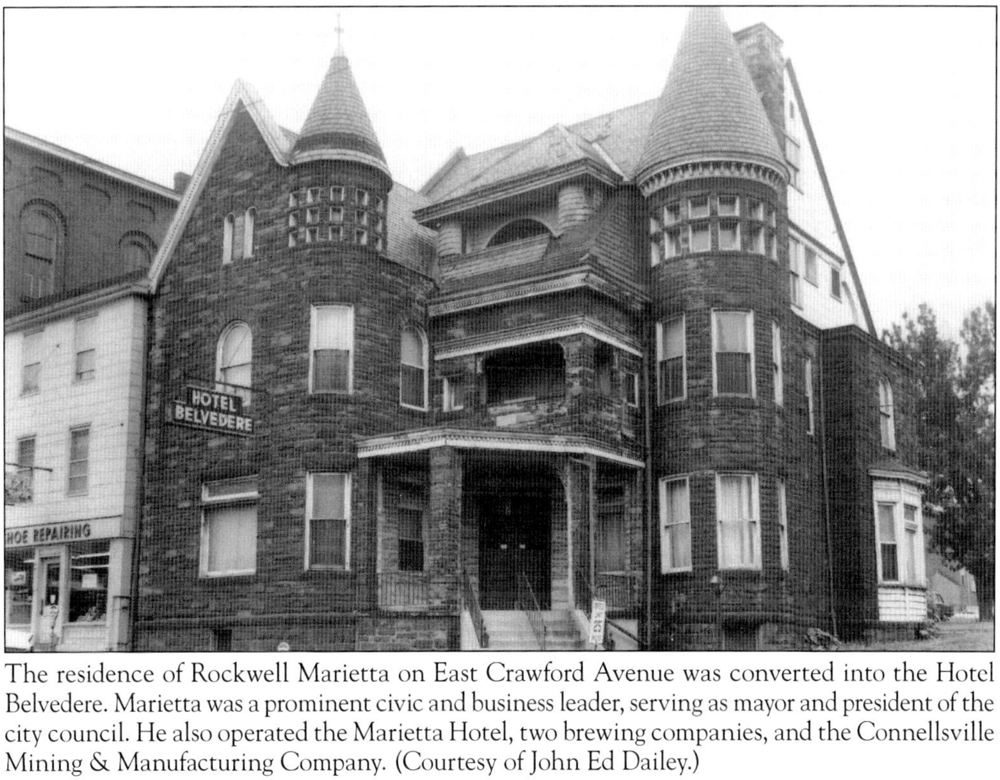

The residence of Rockwell Marietta on East Crawford Avenue was converted into the Hotel Belvedere. Marietta was a prominent civic and business leader, serving as mayor and president of the city council. He also operated the Marietta Hotel, two brewing companies, and the Connellsville Mining & Manufacturing Company. (Courtesy of John Ed Dailey.)

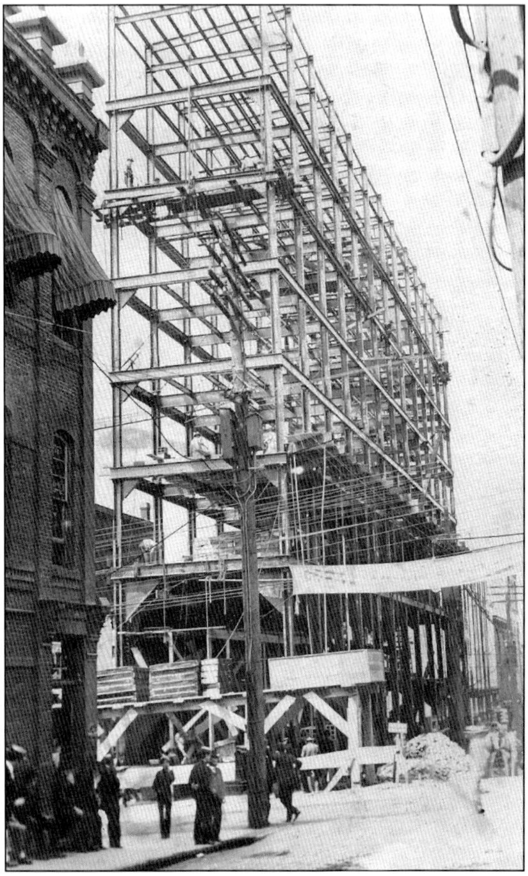

The Second National Bank is seen here under construction around 1910 on Brimstone Corner. The original office, partially visible on the left, was directly across the street from the construction site. Joseph McCormick served as the first president. Directors included John Frisbee, Joseph Soisson, and Charles Davidson. (Courtesy of Tom Korba.)

The Hollywood Shoppe, seen below at night around 1950, was typical of the businesses on West Crawford Avenue, with their bright lights, attractive storefronts, and window displays. This ladies clothing store was started by Louis Barron in 1935 and moved from Pittsburgh Street to this location, next to the Second National Bank, in 1949.

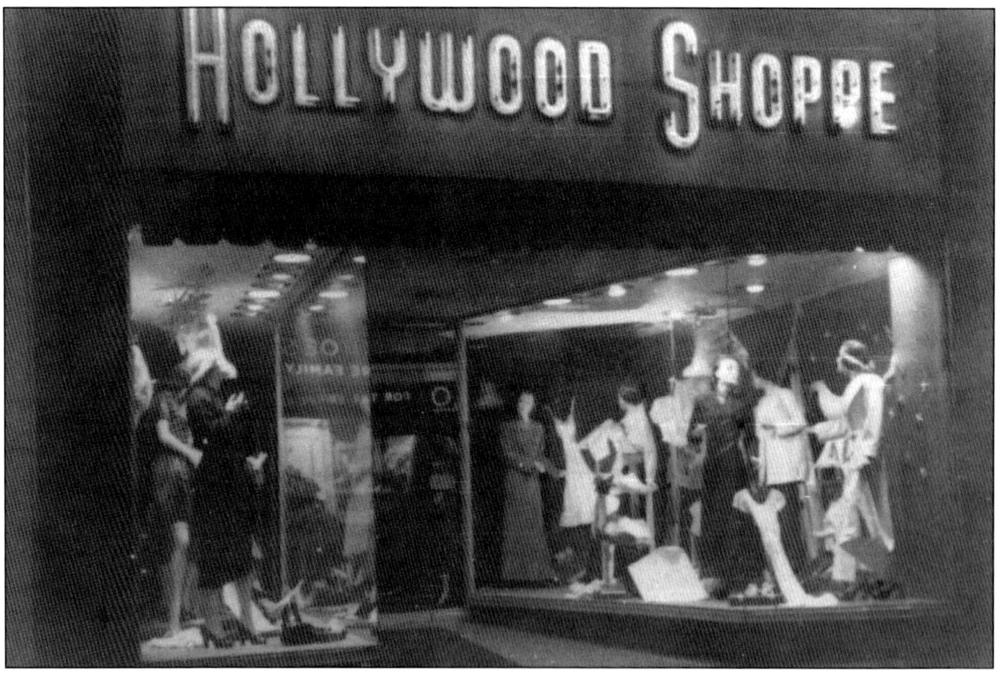

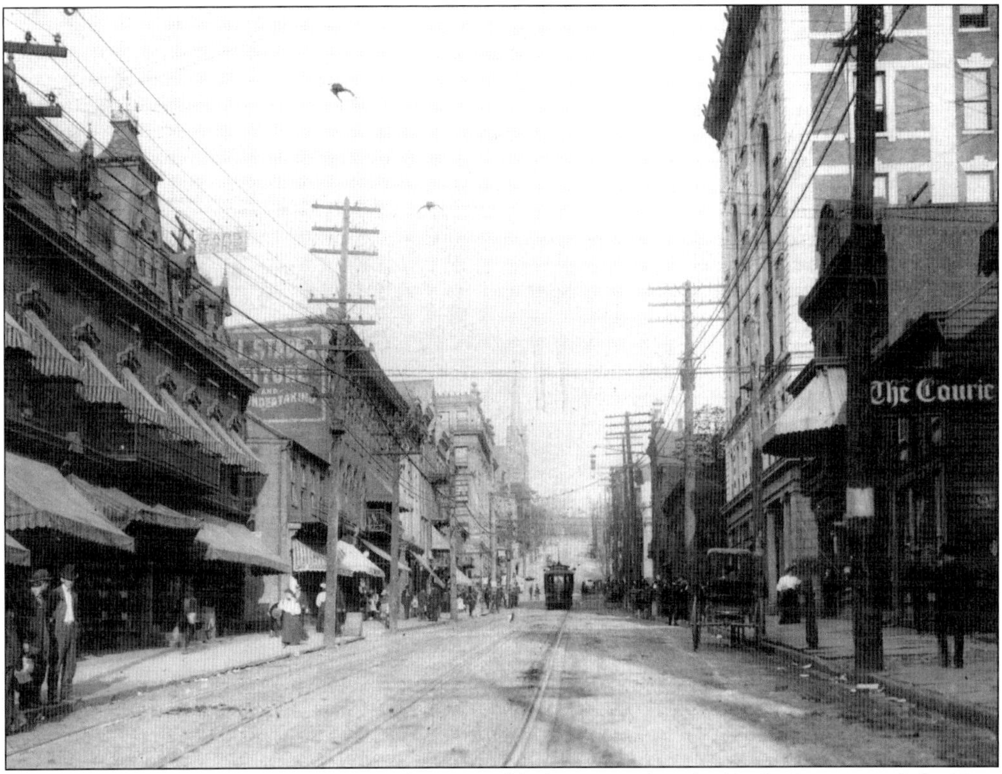

Crawford Avenue is seen above around 1890 and below in 1950. The *Daily Courier* and the First National Bank are on the right above. The New York Clothing Company occupies the building in the left foreground. The Soisson Building is near the center of the same block. Below, a West Penn streetcar passes Sun Drug Store en route to the terminal on Arch Street. The familiar orange transit vehicles were built between 1910 and 1925 with a seating capacity of 60 passengers. In spite of the different time periods, both photographs capture the essence of downtown Connellsville as a leading commercial and financial center. (Below, courtesy of Sandy Russell.)

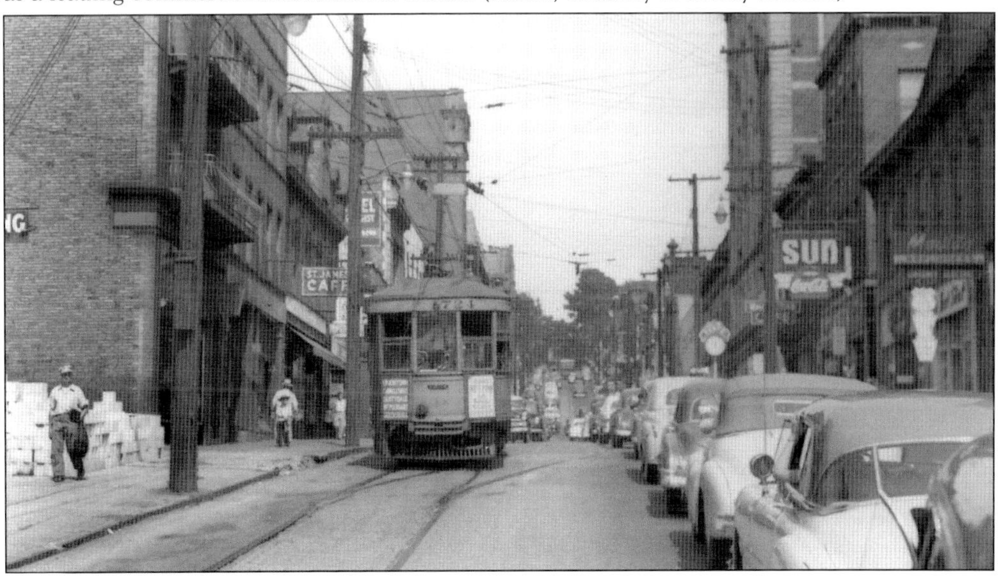

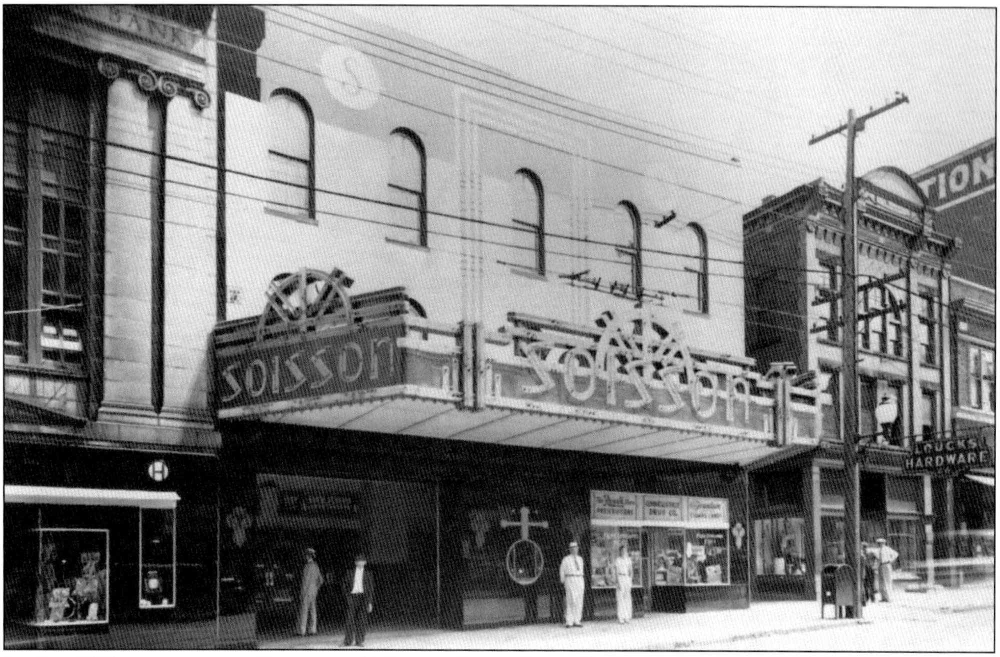

In the 1950s, young people spent all day on Saturday at the Soisson Theatre viewing movies for kids followed by the special feature. The Crawford Avenue theater was established by Joseph Soisson in 1907. It was originally designed for live performances, but management switched to showing films in 1928. It had a seating capacity of 1,000 patrons. (Courtesy of Felix Prestia.)

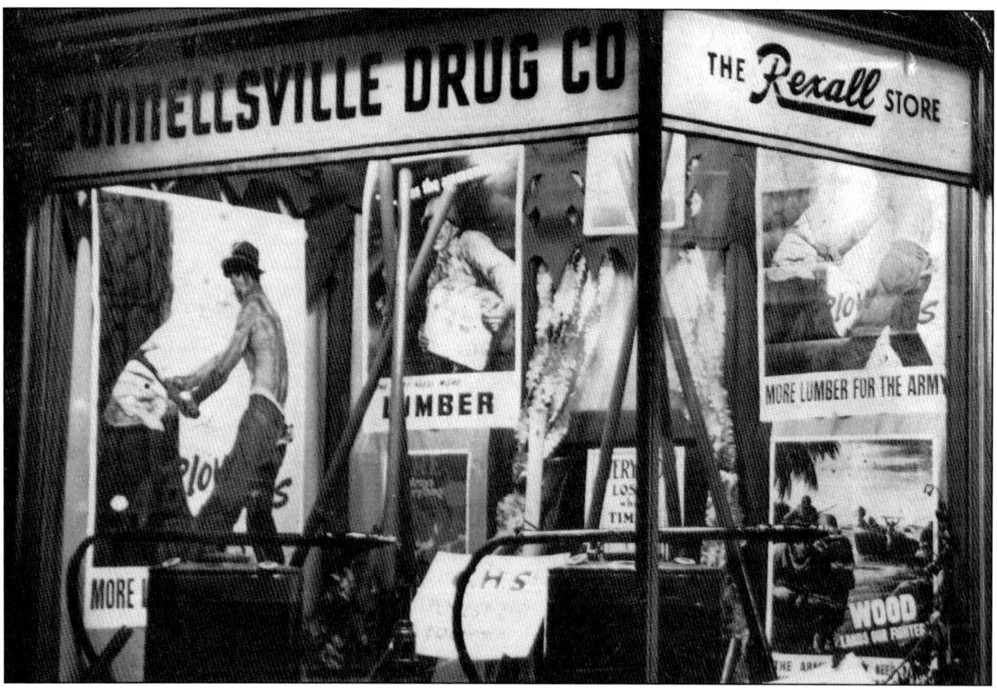

The illuminated window of the Connellsville Drug Co. is seen here during World War II. Note the posters encouraging lumber companies to increase production for the war effort. The store was located in the Soisson Theatre building. (Courtesy of Tom Korba.)

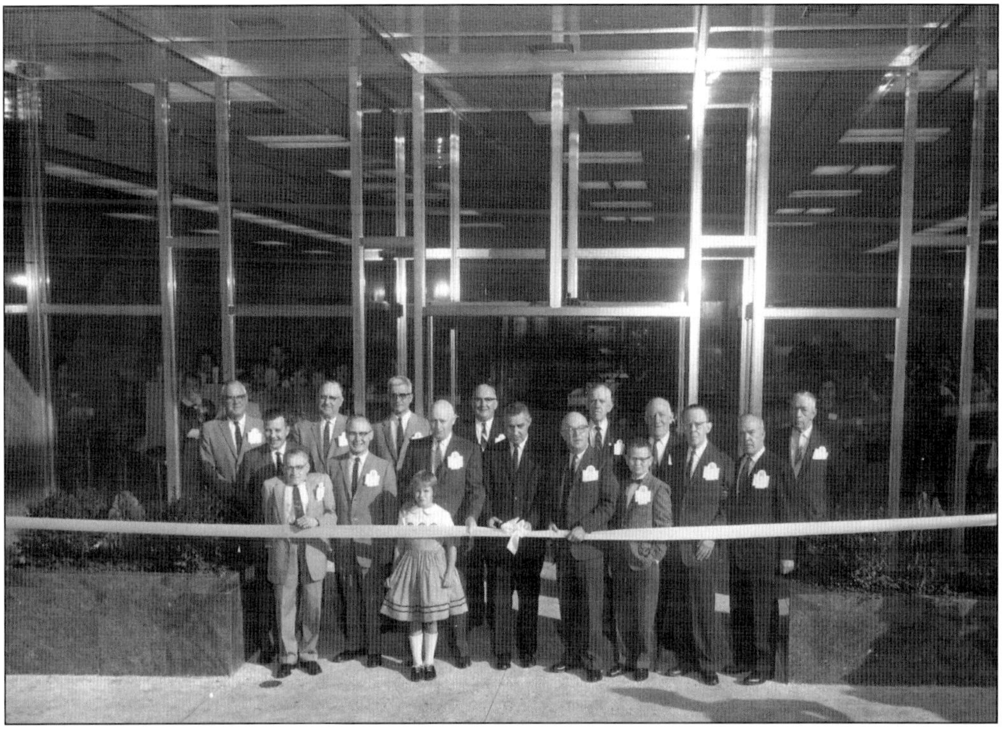

Above, officers and guests pose at the opening of the Gallatin Bank on Crawford Avenue in 1958. They are, from left to right, (first row) Samuel Magie, Lawrence Weaver, Hugh Handford, Polly Morton, Robert McLuckey (mayor), Paul Malone, and Thomas Burkhardt; (second row) Thomas Whyel, J. Lewis Williams, I.N. Hagen, Orville Eberly, Daniel Swaney, Joseph Ray Jr., James Hankins, Meyer Aaron, James Driscoll, and A.W. McDowell. The site is now occupied by PNC Bank.

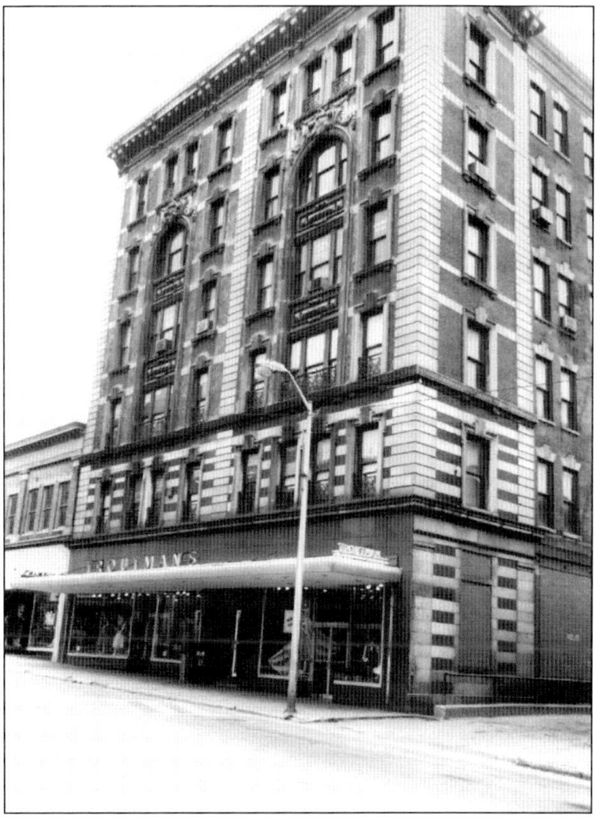

The A.E. Troutman Company was founded in Greensburg and opened this department store in 1925. With three floors of merchandise, it was the largest retail establishment in Connellsville. West Penn streetcars made regular stops in front of the store after leaving the Arch Street terminal. The building was originally constructed in 1903 as the headquarters of First National Bank. (Courtesy of Felix Prestia.)

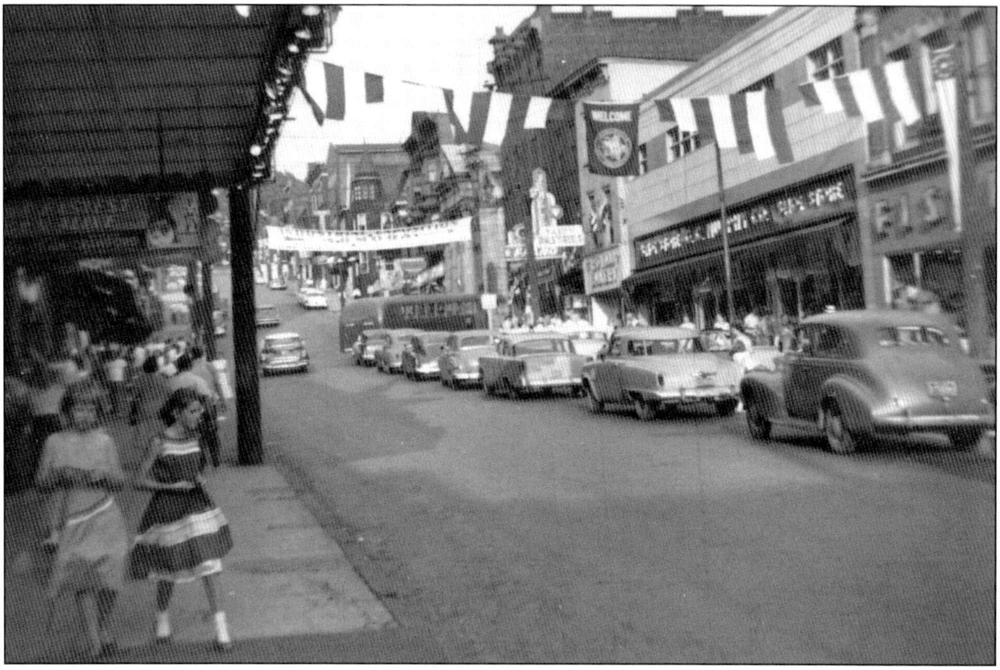

The days of summer attracted large crowds for shopping on Crawford Avenue, as seen in this 1956 image taken under the marquee of the Soisson Theatre, looking east toward Brimstone Corner. The stores on the right, which include Fisher's, G.C. Murphy's five-and-dime store, and the Royal Bakery, are decorated for the town's sesquicentennial. The banner across the street advertises an industrial exposition at the National Guard armory. (Courtesy of Tillie Berg.)

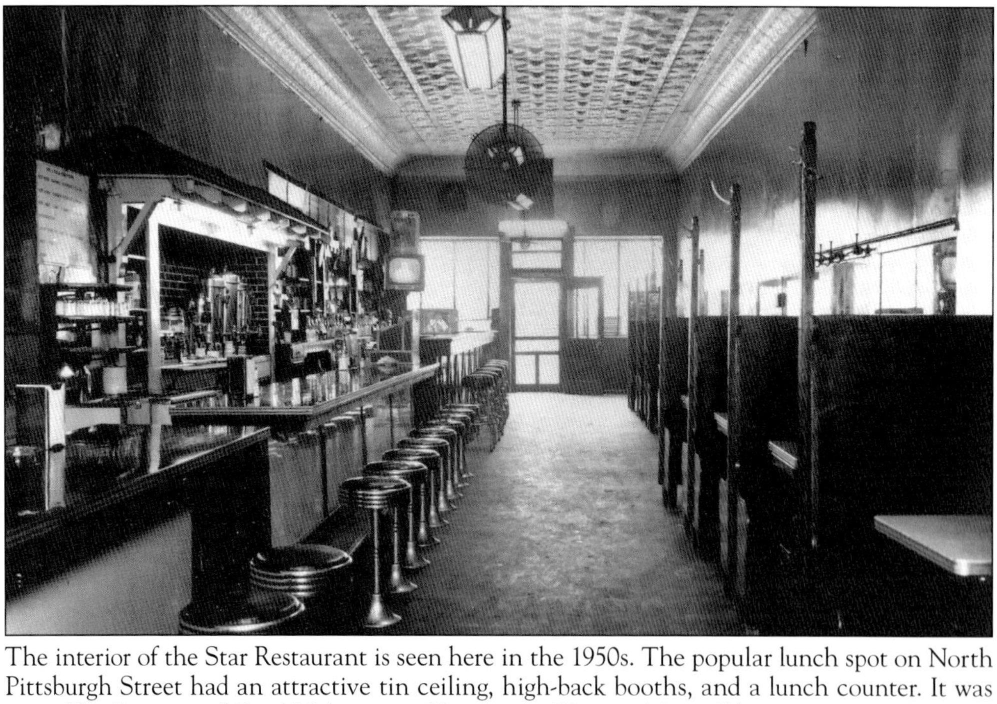

The interior of the Star Restaurant is seen here in the 1950s. The popular lunch spot on North Pittsburgh Street had an attractive tin ceiling, high-back booths, and a lunch counter. It was owned by George and Pearl Melassanos. (Courtesy of Fotenie Mongell.)

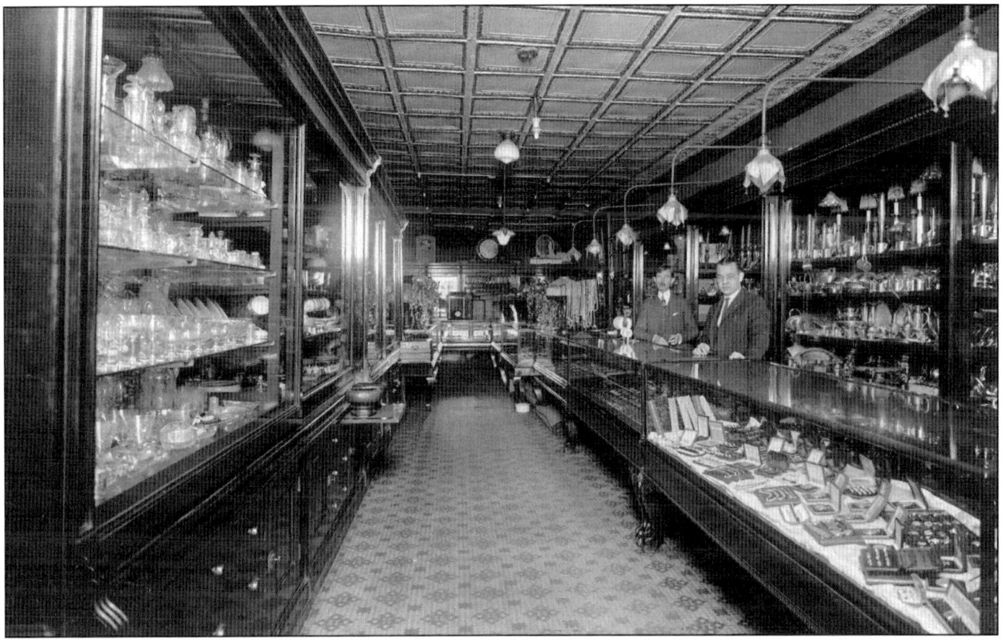

Albert Kurtz (left) and an employee pose inside their jewelry store on West Crawford Avenue. The business was established in 1892 and was located near what is now the new Connellsville Canteen. Note the lighting fixtures and the fine array of merchandise. Kurtz's also had a contract with the B&O Railroad for calibrating watches. Engineers and conductors were required to have their watches serviced there. (Courtesy of Joseph Kurtz.)

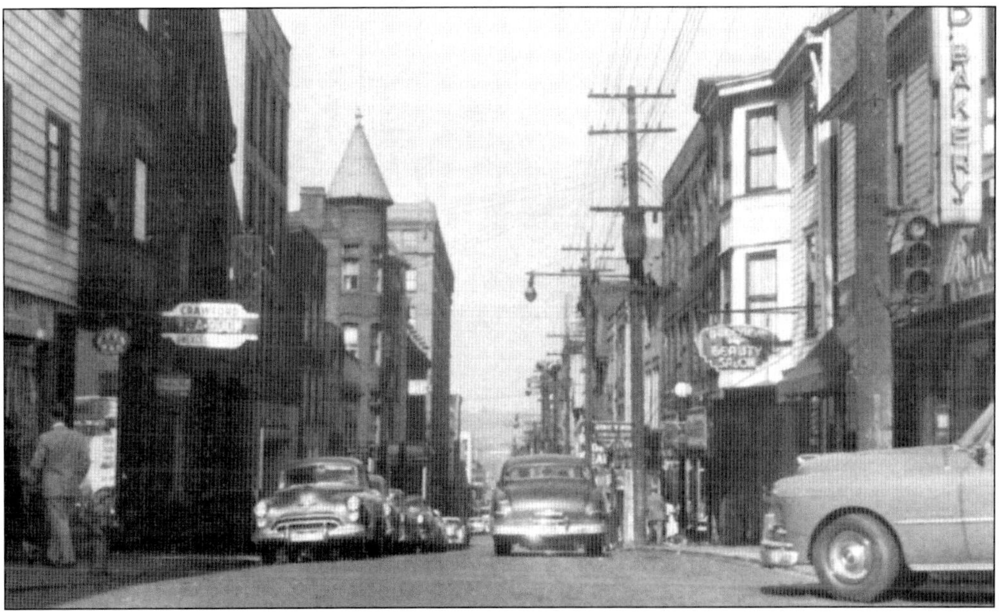

The busy intersection of South Pittsburgh Street and Fairview Avenue is seen here in the 1950s. A sign for the Crawford Tea Room is visible on the left. This popular restaurant was adjacent to the former residence of Dr. J.C. McClanathan. His home was listed in the National Register of Historic Places in 2002. Also, the former Willard Hotel, with its turret and bell-shaped crown, is near the middle of the block. On the right, in the foreground, is the Double B Bakery.

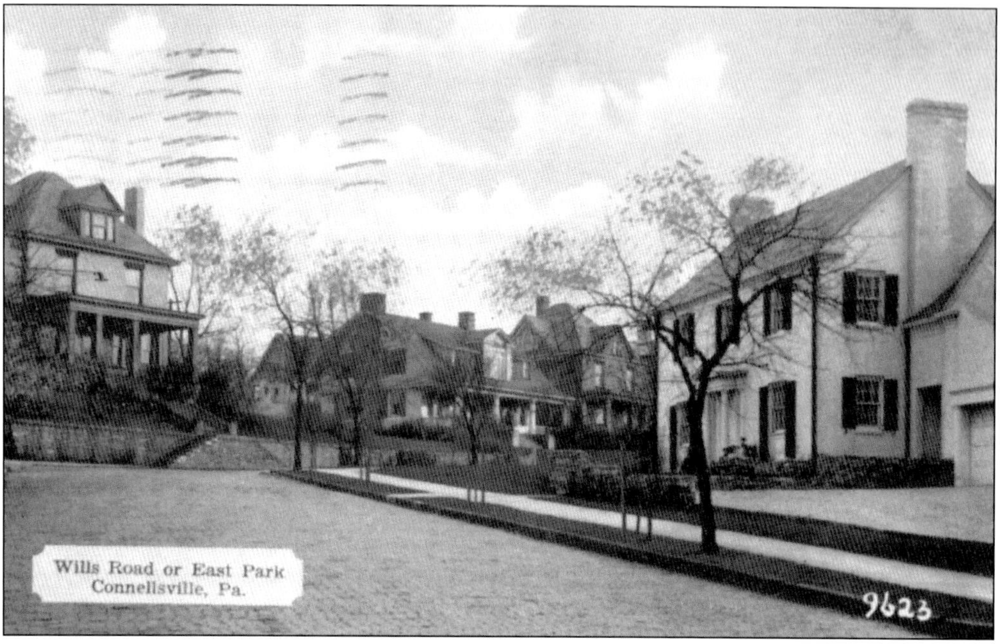

Prominent businessmen and professionals built these stately homes on Wills Road, as seen above from Angle Street. The road was named for Will, the brother of James Hogg, who owned extensive property in Connellsville and worked as a civil engineer for the H.C. Frick Coke Company. The image below illustrates the bucolic character of the road around 1900, when it provided an attractive setting for carriage rides and other leisure activities. (Both, courtesy of Barbara and Tony Keefer.)

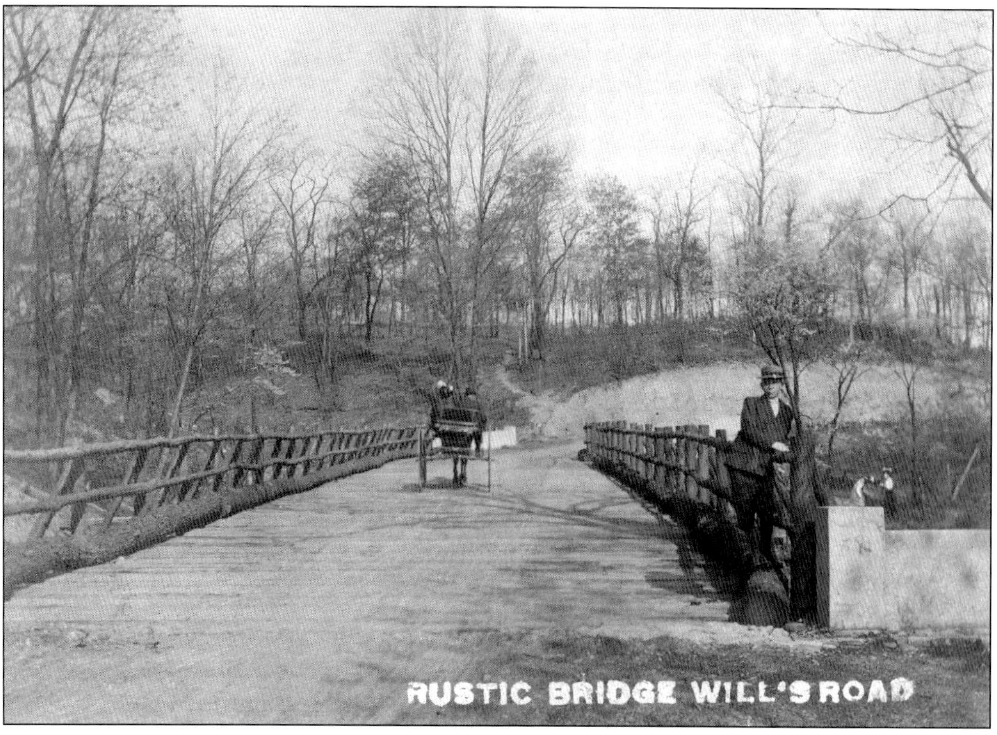

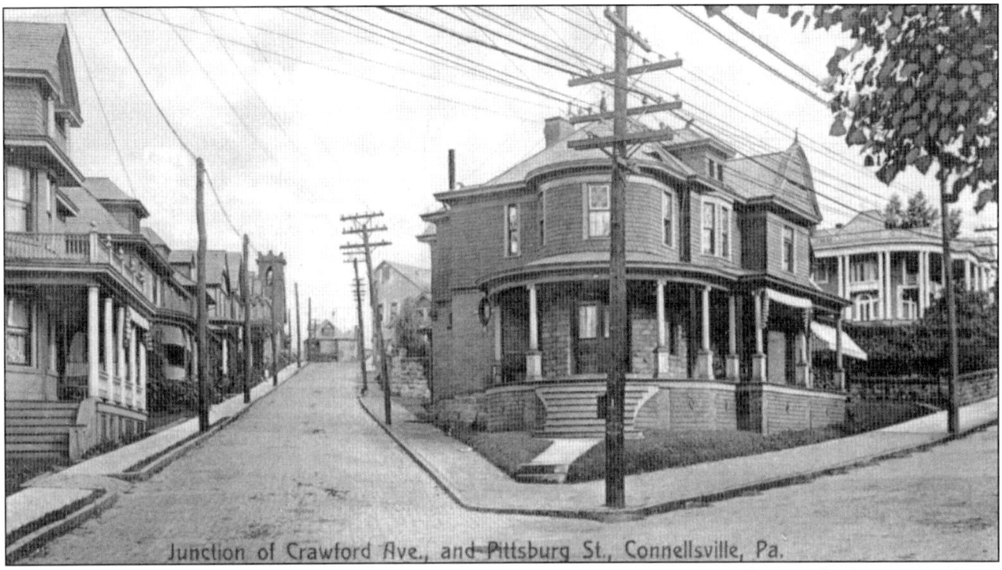

This uniquely shaped house on the corner of South Pittsburgh Street and Crawford Avenue (now Lincoln Avenue) is seen here in 1900. The Linford Ruth house is partially visible on the right. The church in the left background is Otterbein Methodist Church. (Courtesy of Sandy Russell.)

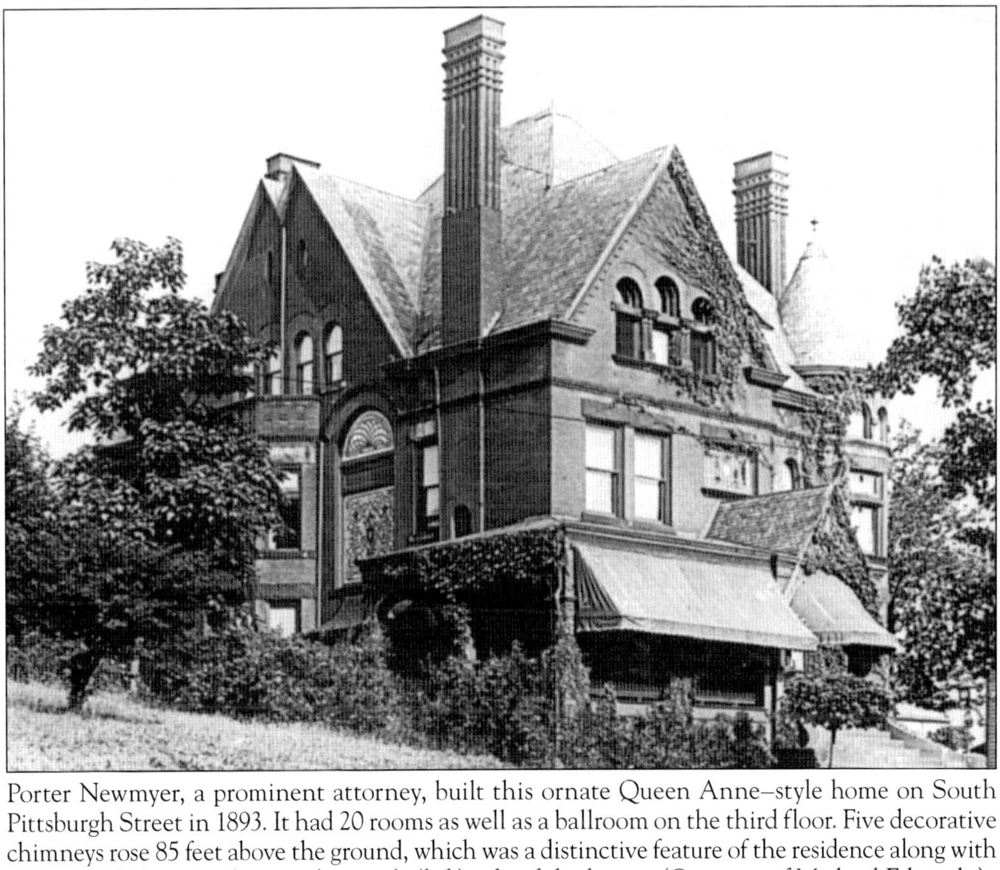

Porter Newmyer, a prominent attorney, built this ornate Queen Anne–style home on South Pittsburgh Street in 1893. It had 20 rooms as well as a ballroom on the third floor. Five decorative chimneys rose 85 feet above the ground, which was a distinctive feature of the residence along with the stained-glass window on the north (left) side of the house. (Courtesy of Michael Edwards.)

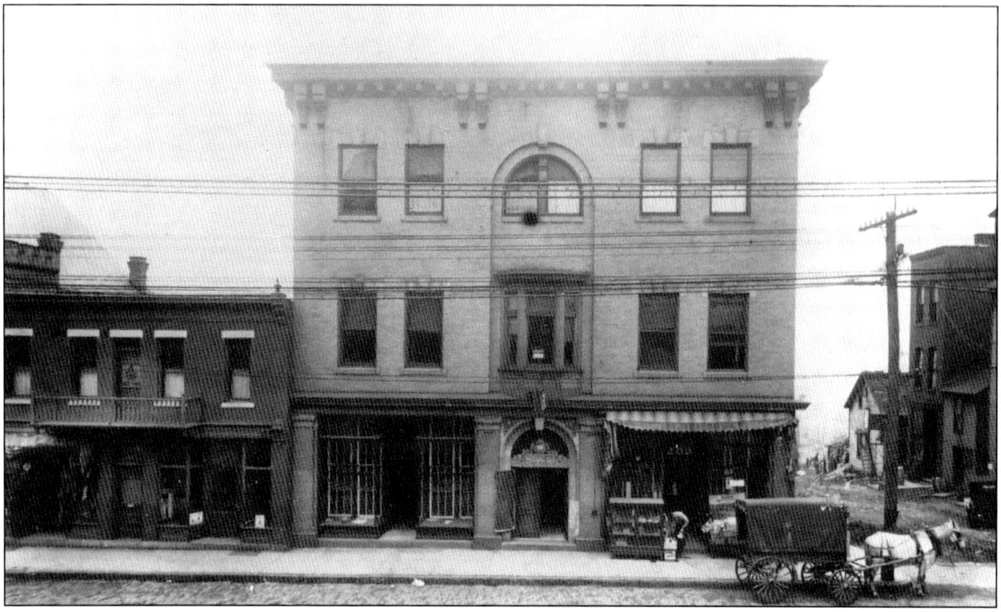

The Connellsville YMCA was formed in 1893 and occupied a series of buildings before this facility was constructed in 1906 on the corner of South Pittsburgh Street and West Fairview Avenue. In addition to meeting rooms and a restaurant, it included a swimming pool and a gymnasium in the basement. The YMCA also provided lodging for railroad employees who had layovers between assignments in Connellsville.

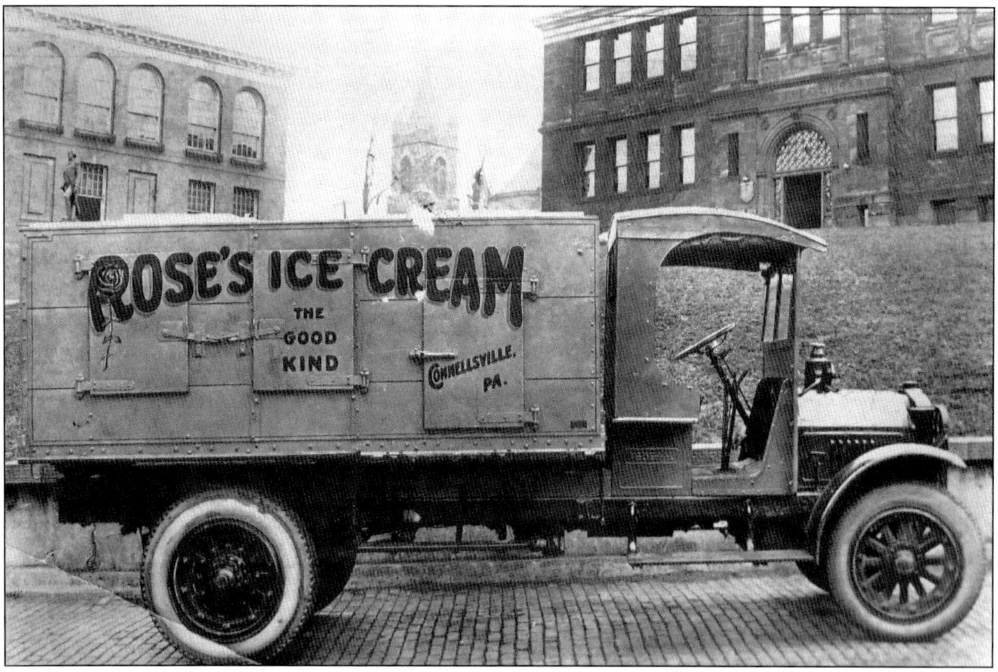

Rose's Ice Cream Company delivered their products in this truck in the 1920s. The enterprise was founded by F.C. Rose in 1898 and continued as a Connellsville business until 1982. It was originally located at 408 North Pittsburgh Street and moved to McCormick Avenue in 1932. This property is now occupied by the Rose Square town houses.

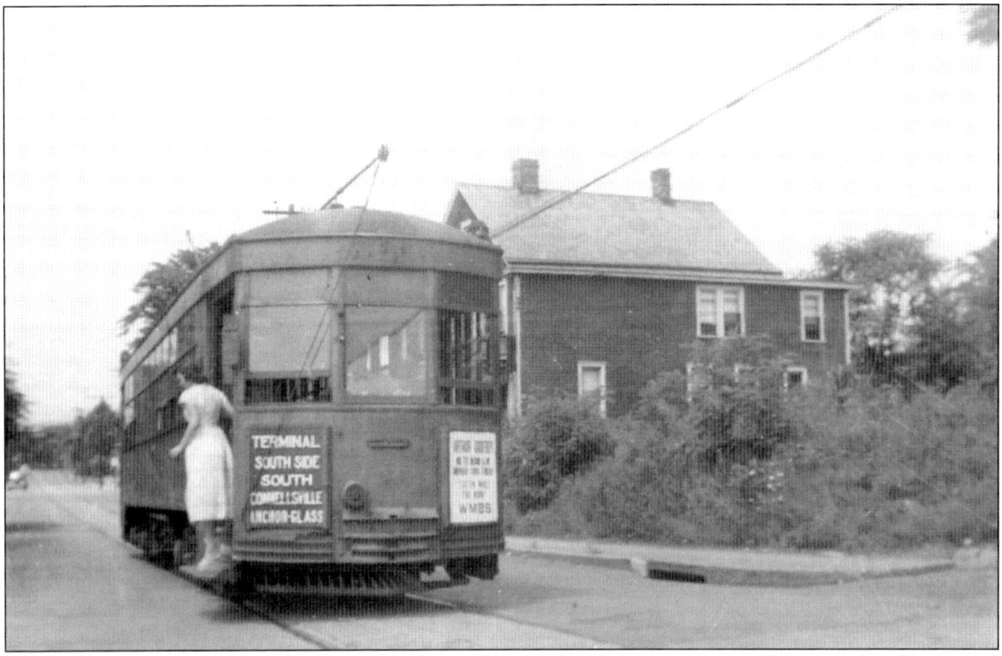

A passenger exits the streetcar at South Pittsburgh and Wine Streets in South Connellsville around 1950. The car is a West Penn 200 series, which were used on the line between the terminal and South Connellsville, a distance of 2.2 miles. They also connected Connellsville with Dickerson Run via Leisenring. (Courtesy of Tillie Berg.)

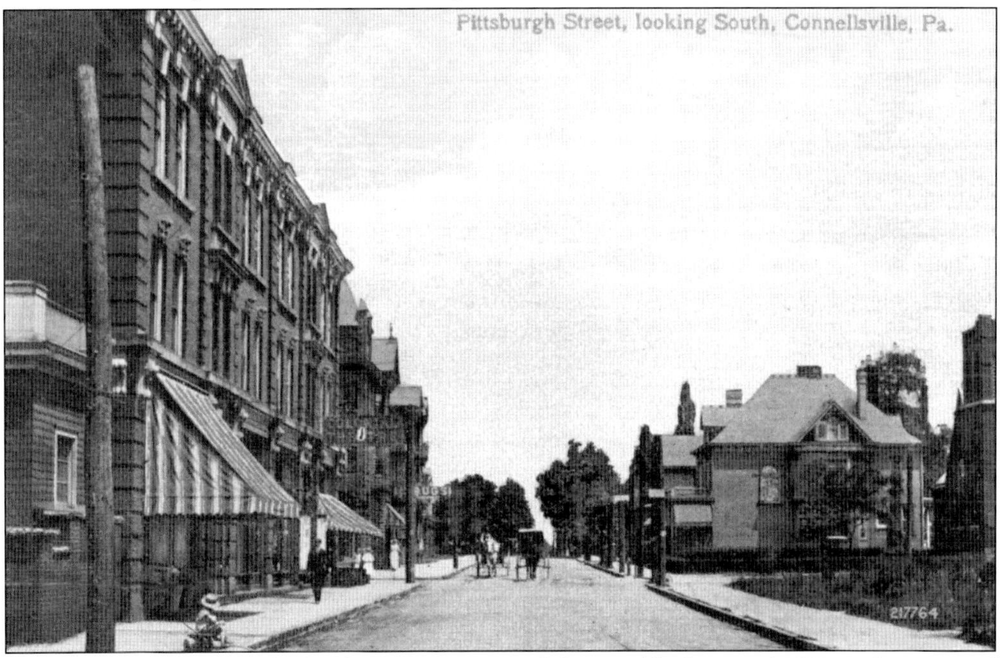

The Colonial Theatre and other businesses occupied this city block around 1910 on the corner of South Pittsburgh and Green Streets. Across Green Street is the South Windsor Building. Keagy's Drug Store occupied the ground floor, and apartments were on the upper two levels. The first structure on the right is the Keagy family residence. (Courtesy of Margaret Molinaro.)

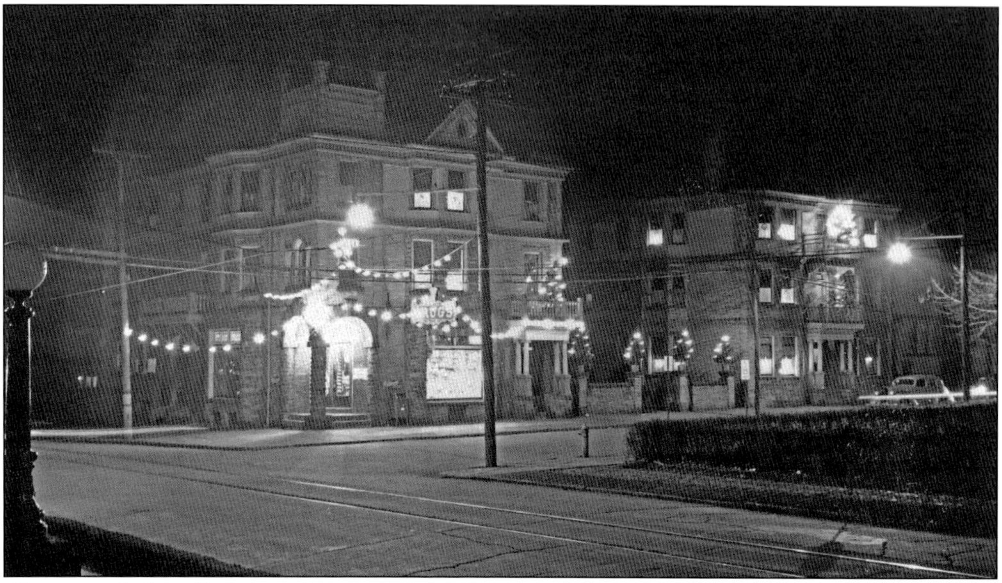

Christmas lights illuminated Keagy's Drug Store and the Windsor Apartments in the 1950s. (Courtesy of Bill Sechler.)

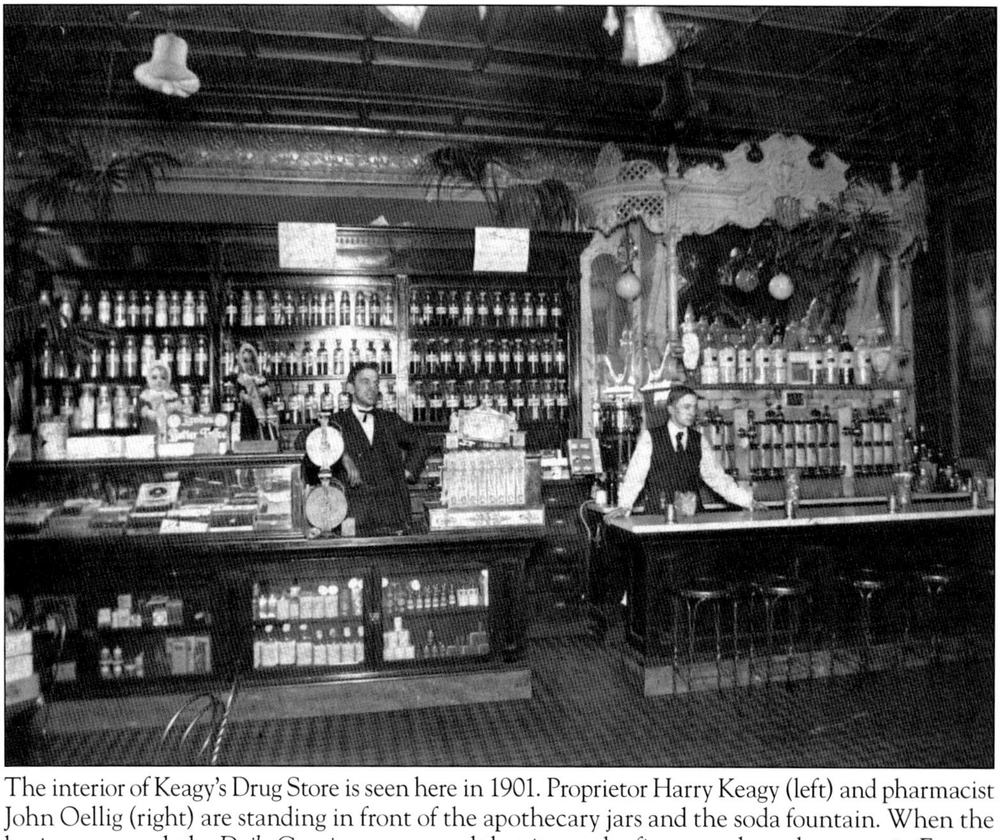

The interior of Keagy's Drug Store is seen here in 1901. Proprietor Harry Keagy (left) and pharmacist John Oellig (right) are standing in front of the apothecary jars and the soda fountain. When the business opened, the *Daily Courier* announced that it was the finest modern pharmacy in Fayette County. (Courtesy of Bill Sechler.)

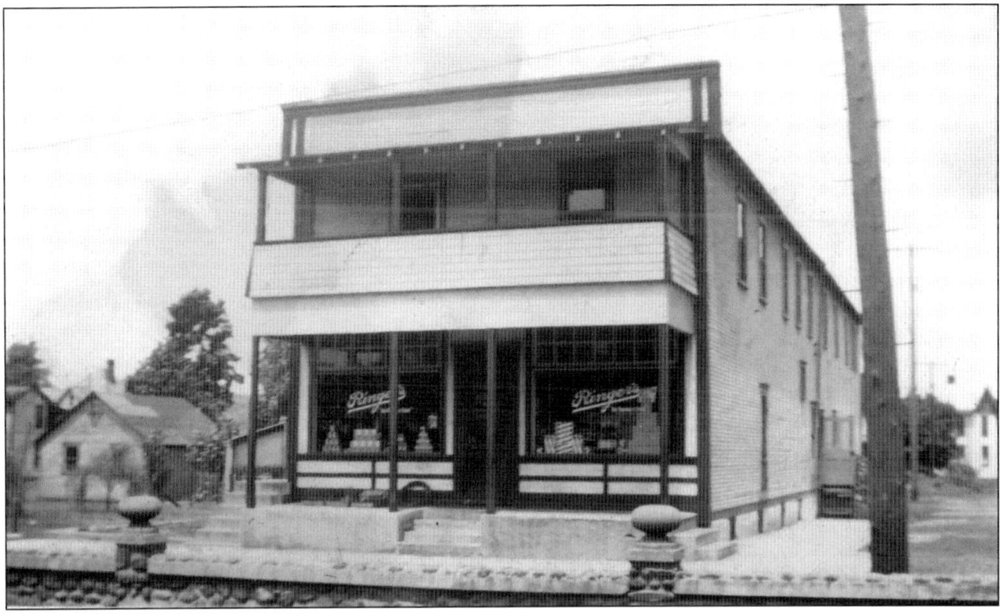

Seen above is Ringer's store, one of several businesses in South Connellsville that sold meat and groceries. It was located on Pine Street and established by William Ringer in 1924. Ringer also operated a lumberyard and used some of his lumber to build the store. It is currently occupied by Fayette Appliances. Below is the Ringer family residence at 111 Pine Street, which dates from the mid-1800s. It was the first home in the borough with indoor plumbing. Reportedly, the stone fence in the foreground was constructed by men who worked on the stone garden at St. Rita's Catholic Church. (Both, courtesy of Pete Maricondi.)

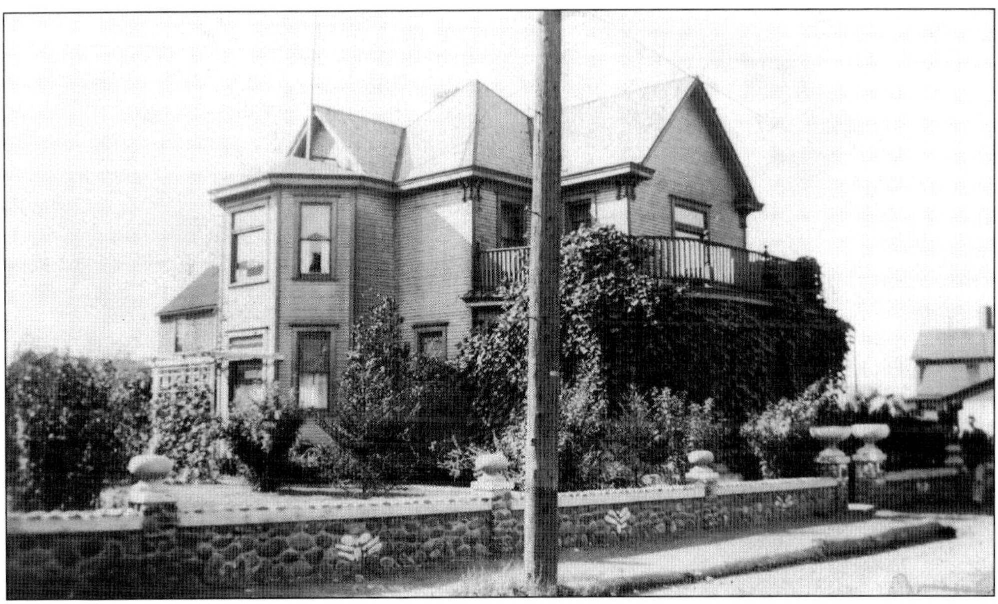

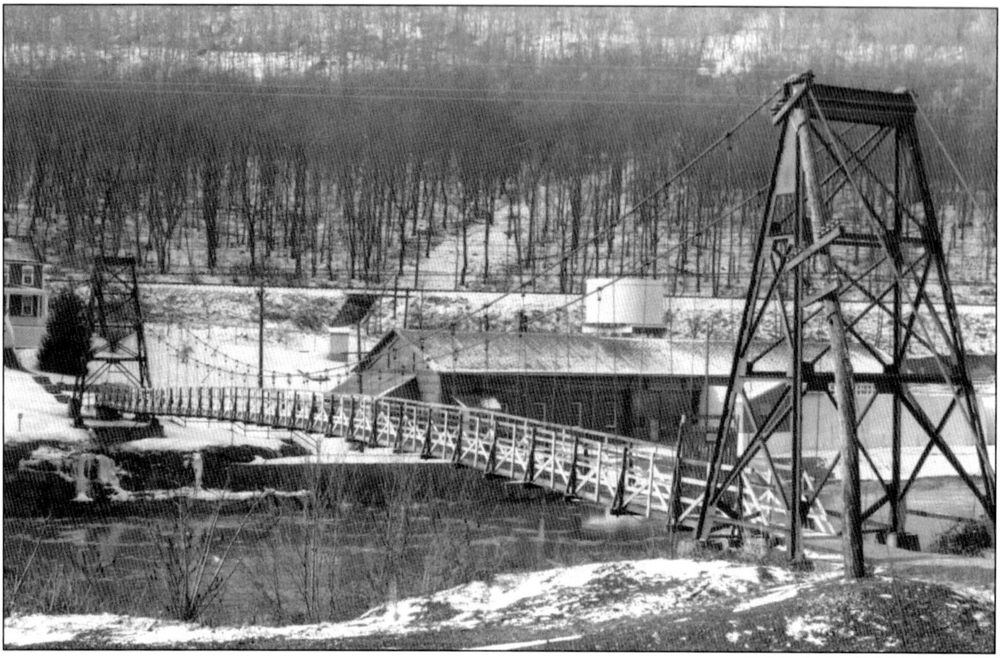

The "swinging bridge" in South Connellsville was built in 1898. It was used primarily to reach a pumping station from the east bank of the Youghiogheny River. The station was part of the Trotter Water Company, which was organized by the H.C. Frick Coke Company. It supplied the water required to manufacture coke at Trotter and at the three Leisenring mines.

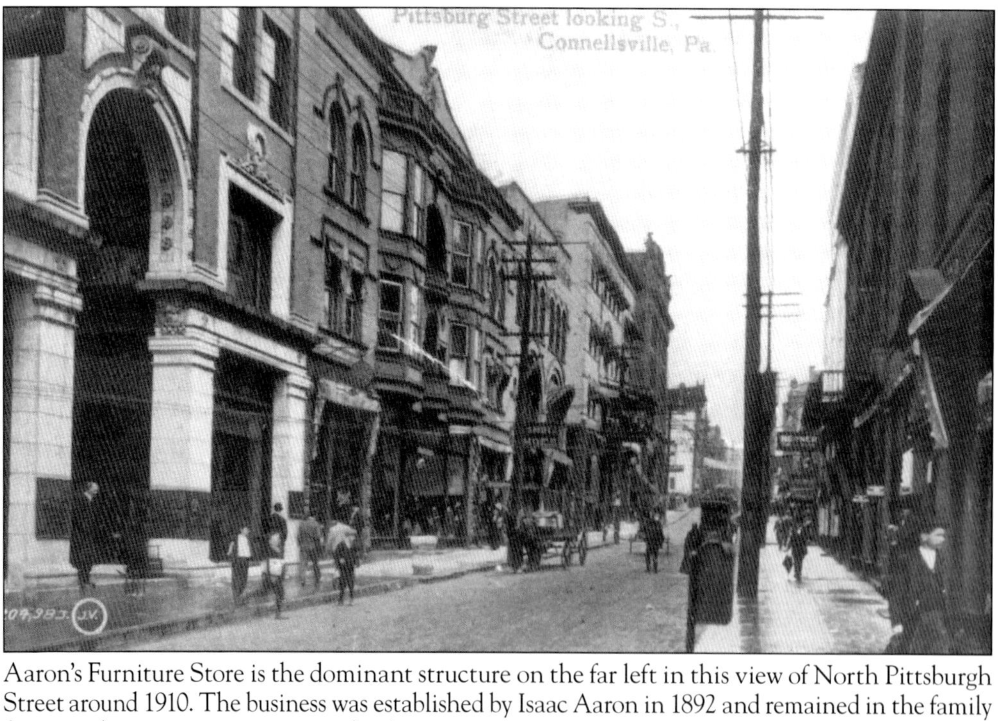

Aaron's Furniture Store is the dominant structure on the far left in this view of North Pittsburgh Street around 1910. The business was established by Isaac Aaron in 1892 and remained in the family for two subsequent generations under the management of Meyer and Arthur Aaron. Meyer Aaron was a trustee of the Gallatin Bank and the treasurer of the Connellsville State Hospital.

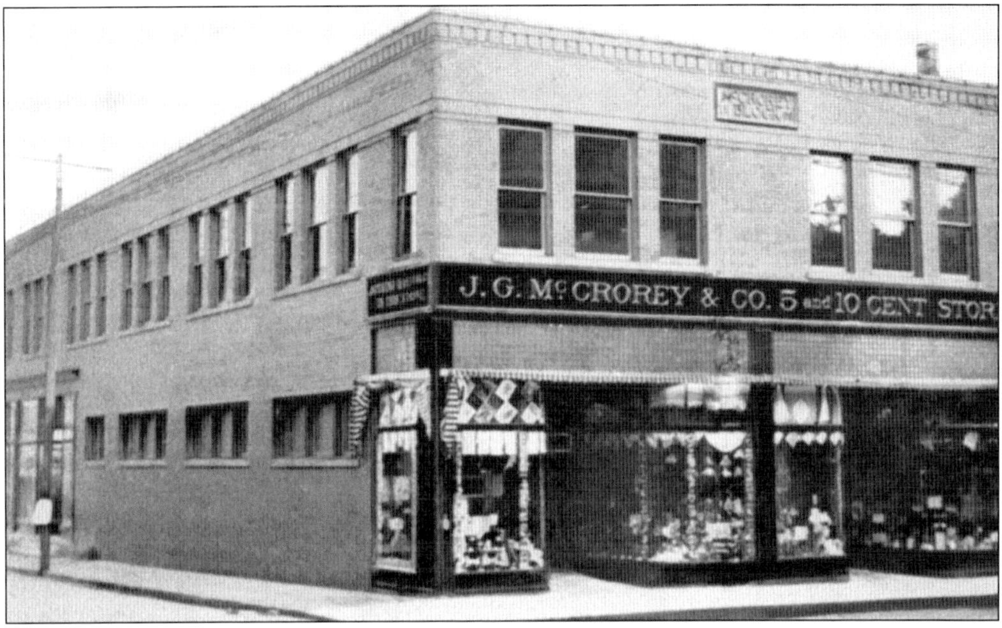

The first McCrorey (Crory) five-and-dime store in Connellsville was located on the corner of North Pittsburgh and Apple Streets. In 1911, there was a gas explosion in the store. Five customers were killed, including Francesco Stirone of Trotter. Stirone saved a girl in the store and returned to help other customers. He did not survive. The fire destroyed six buildings and injured 29 people. (Courtesy of Sandy Russell.)

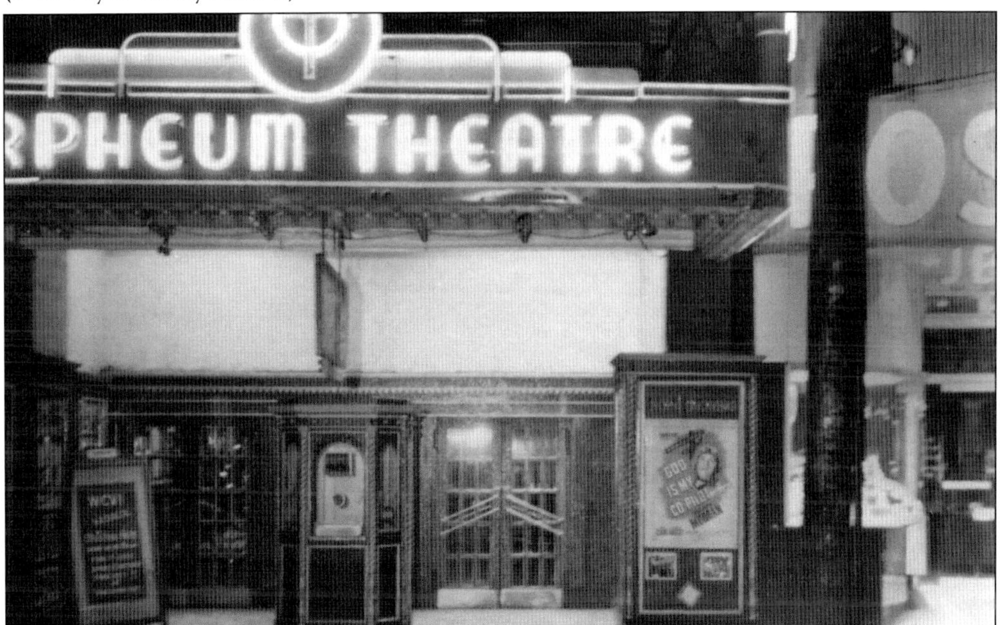

Dennis Morgan was starring in *God is My Co-Pilot* when this photograph was taken of the Orpheum Theatre on North Pittsburgh Street in the 1940s. The entertainment venue opened in 1916 and showed its last picture in 1969. The Woodward family was the final owner. Posner's Jewelry Store is partially visible on the right.

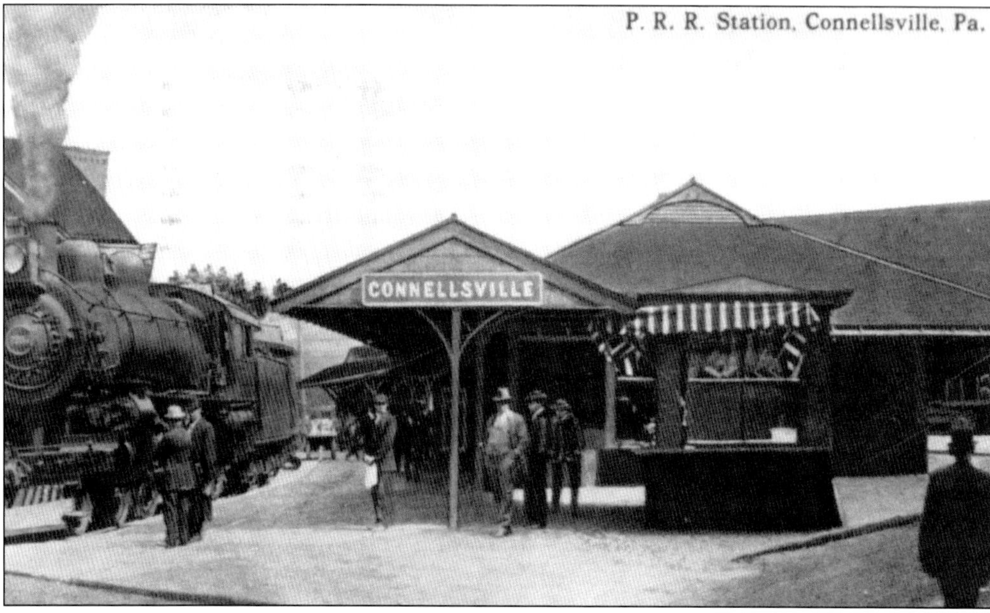

The Pennsylvania Railroad (PRR) reached Connellsville from Greensburg in 1875 over tracks originally laid by the Southwest Pennsylvania Railroad. The Southwest was incorporated by Daniel Rogers Davidson and partners from Westmoreland County. The PRR passenger depot is seen here in 1906. It was on North Pittsburgh Street near Liberty Square. (Courtesy of Felix Prestia.)

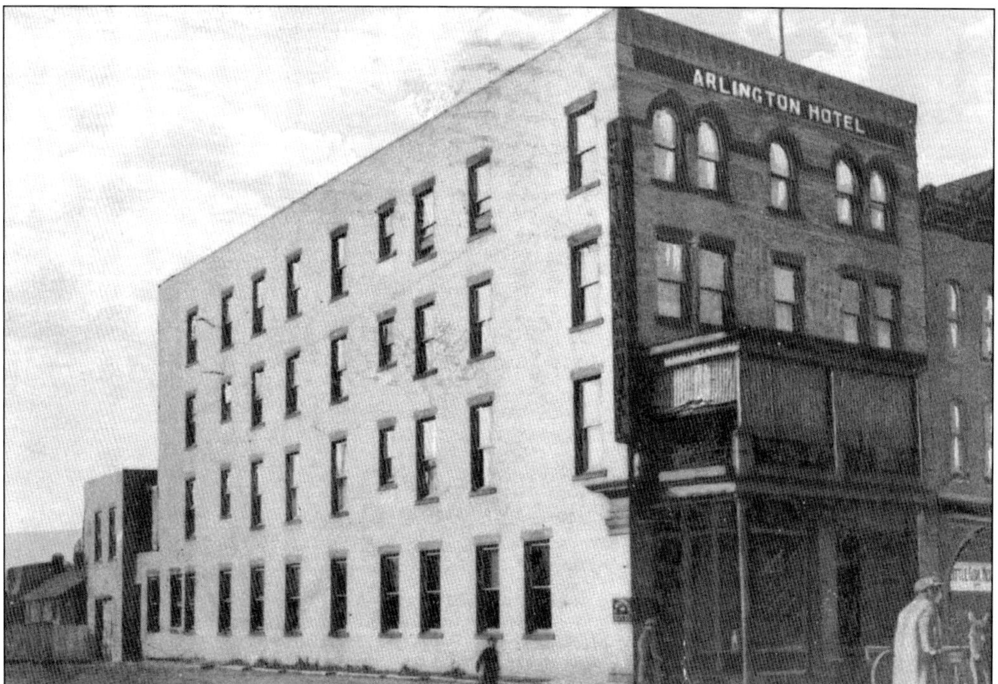

In the early 1900s, Connellsville had eight hotels. Initially, they were concentrated on Water Street near the B&O passenger depot. Others were subsequently established on Crawford Avenue and South Pittsburgh Street. Seen here is the Arlington Hotel, conveniently located near the PRR passenger station. (Courtesy of Barbara and Tony Keefer.)

The *Connellsville Herald*, founded in 1815, was the first local newspaper. Others followed, including the *Enterprise* and the *Keystone Courier*. The first daily paper was the *Connellsville News*. It was printed in this office on the corner of West Apple and North Arch Streets.

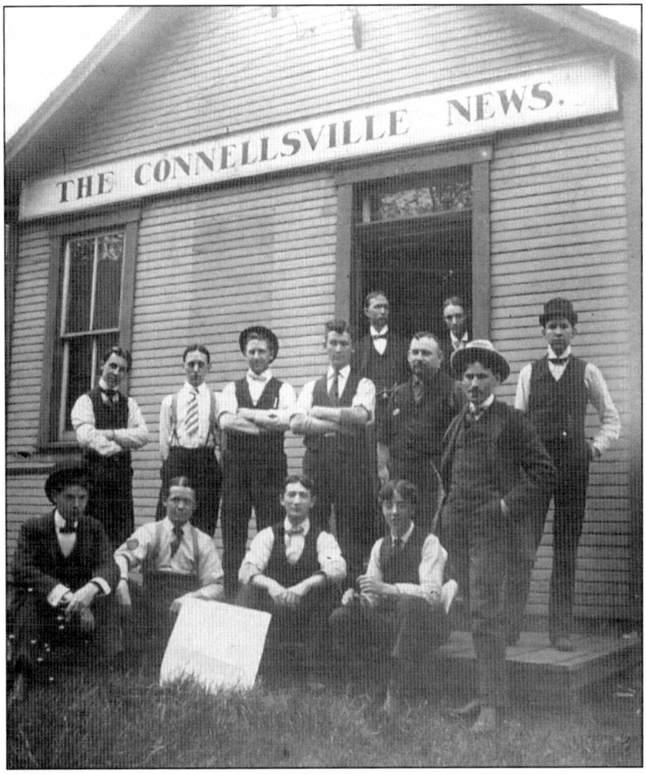

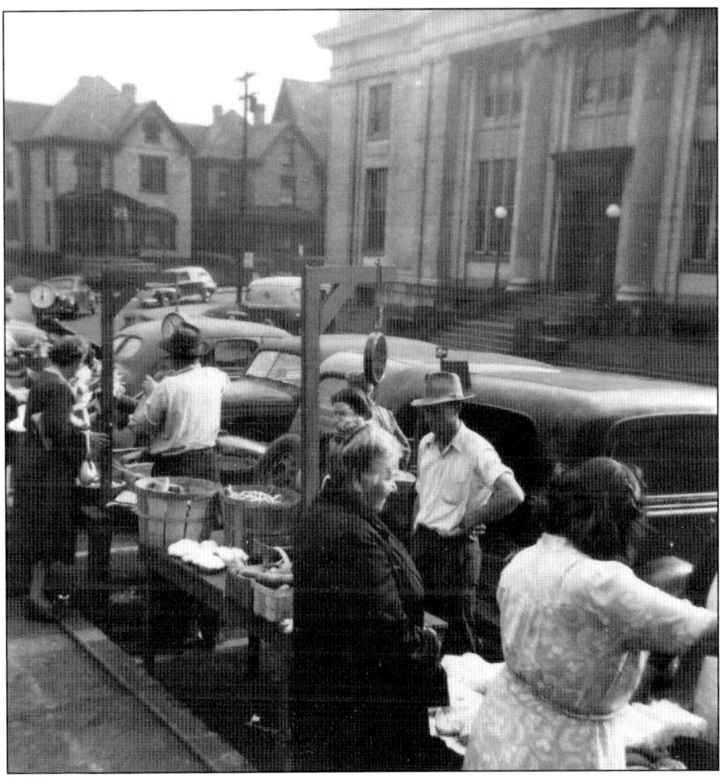

Although Connellsville was a large manufacturing center, it was surrounded by prosperous farms that offered fresh produce to "town folk" in various locations throughout the city. This photograph shows a farmer's market on North Arch Street, across from the post office, around 1940. (Courtesy of John Ed Dailey.)

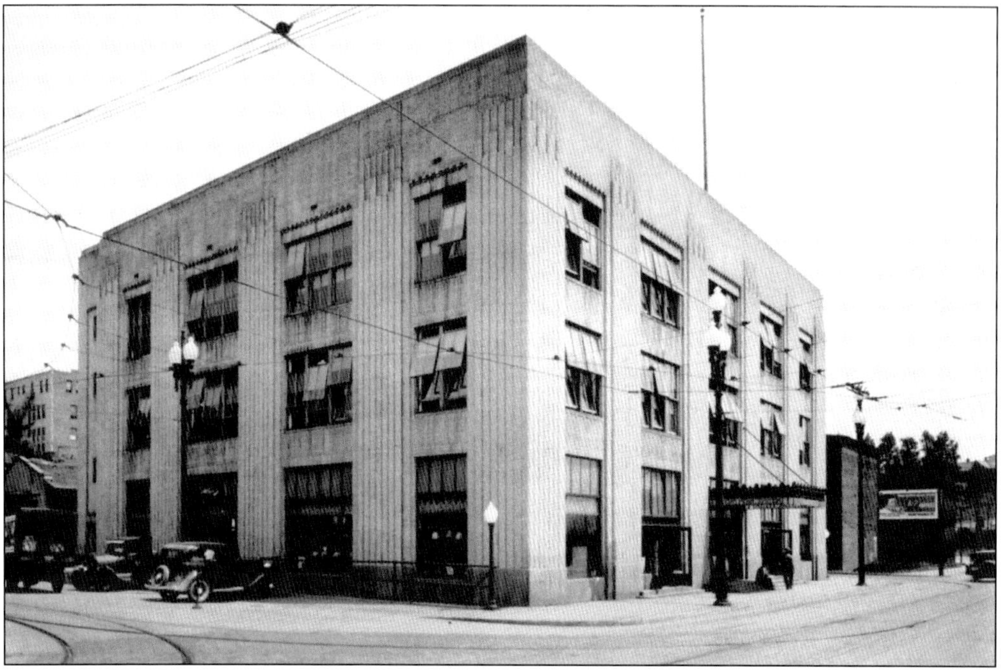

The West Penn Railways terminal (above) was built in 1928 on the corner of West Fairview Avenue and South Arch Street. Also built in that year was a freight station, which is partially visible to the far left, directly behind the terminal. The terminal served as the headquarters of the rail company in addition to being a major transfer point on the main line from Greensburg to Uniontown. The building is currently occupied by the Scottdale Bank & Trust Company. The property is seen below in 1920 before the terminal was constructed. The large building on the left was a factory that manufactured terra-cotta pipe. The upper levels of the First National Bank are partially visible behind the factory. (Above, courtesy of Scottdale Bank & Trust Company; below, courtesy of Tom Rusnack.)

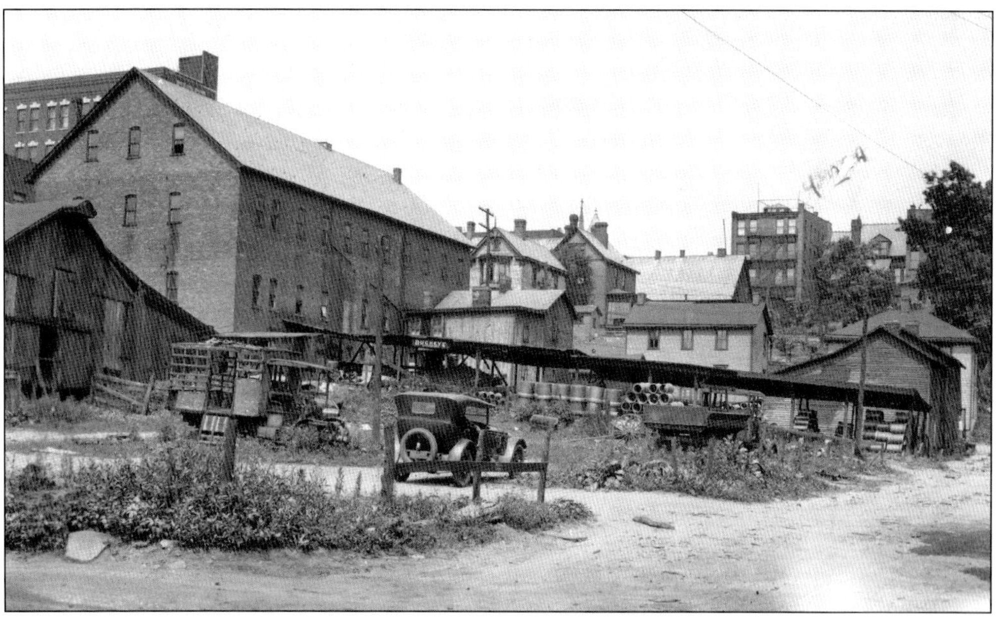

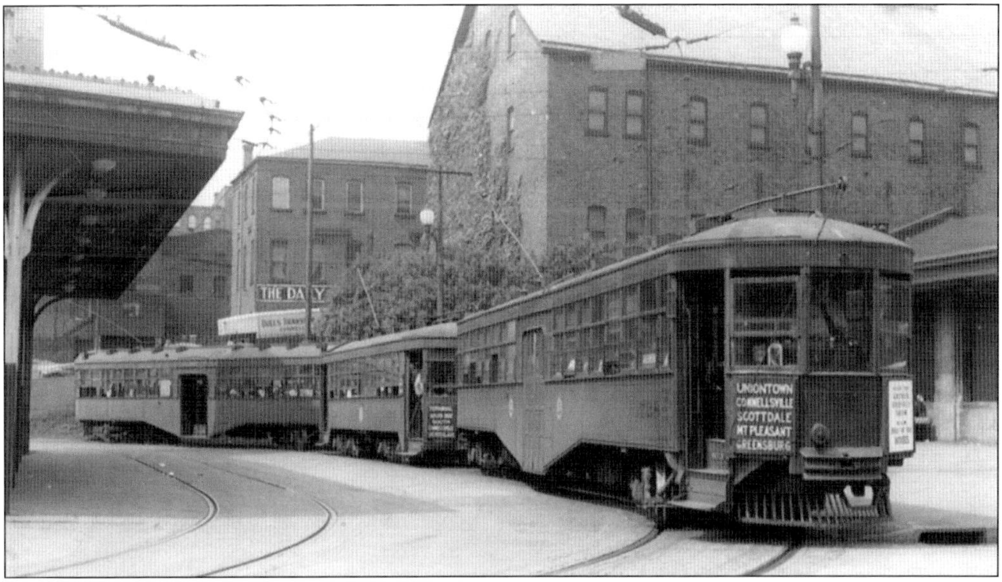

These streetcars make their scheduled connections at the Connellsville terminal in this 1950s image. The first car will depart for Uniontown and the other cars will travel to South Connellsville and Greensburg. According to William Volkmer in his 1999 book *Pennsylvania Trolleys*, Connellsville was the last great interurban hub in America. (Courtesy Scottdale Bank & Trust Company).

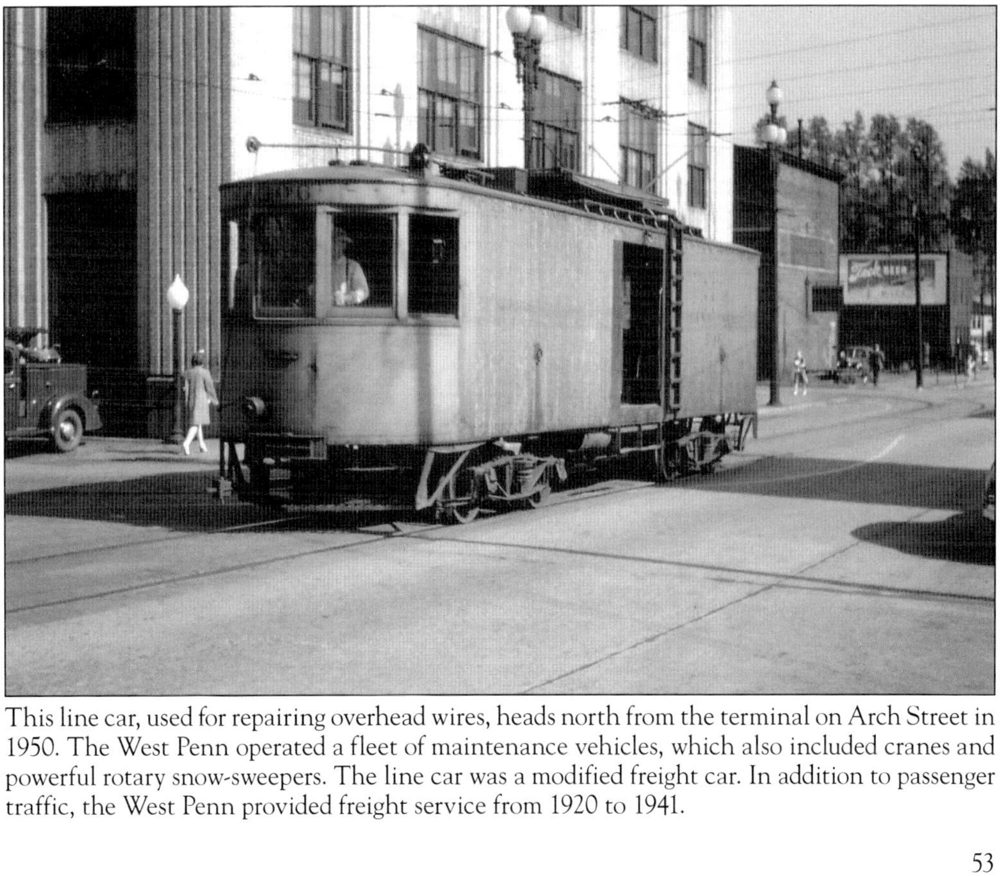

This line car, used for repairing overhead wires, heads north from the terminal on Arch Street in 1950. The West Penn operated a fleet of maintenance vehicles, which also included cranes and powerful rotary snow-sweepers. The line car was a modified freight car. In addition to passenger traffic, the West Penn provided freight service from 1920 to 1941.

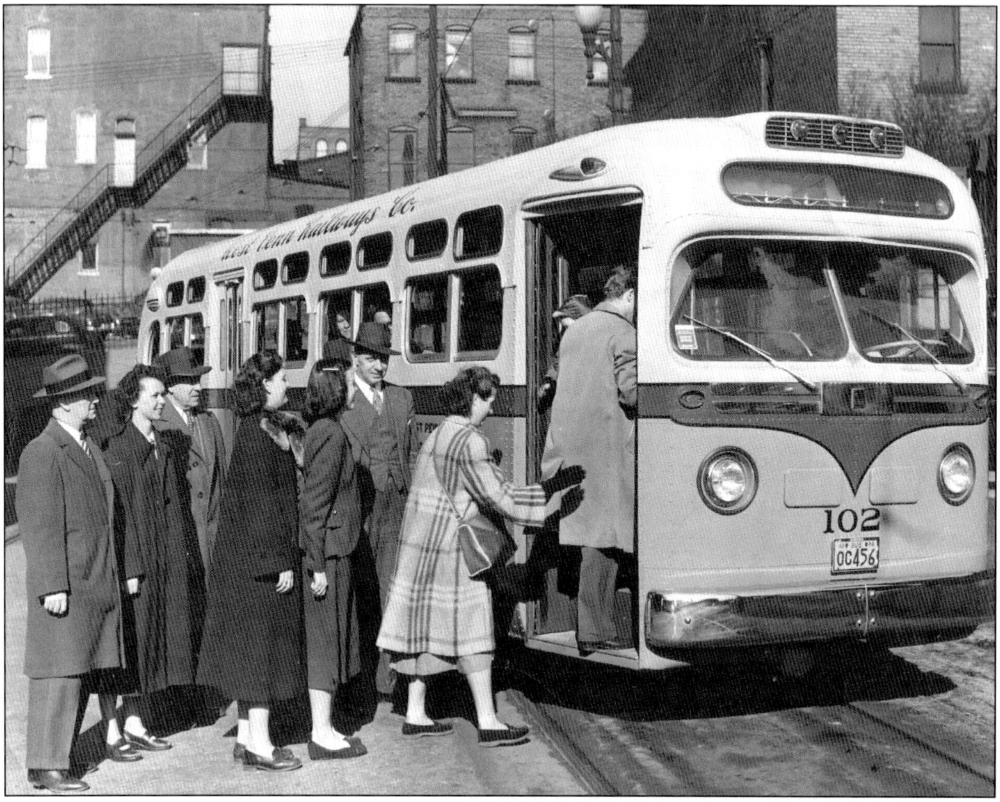

Due to changes in the travel habits of Americans and a loss of population in the West Penn service area, West Penn replaced trolleys with buses which operated between 1951 and 1958. Here, passengers board one of the buses at the Connellsville terminal.

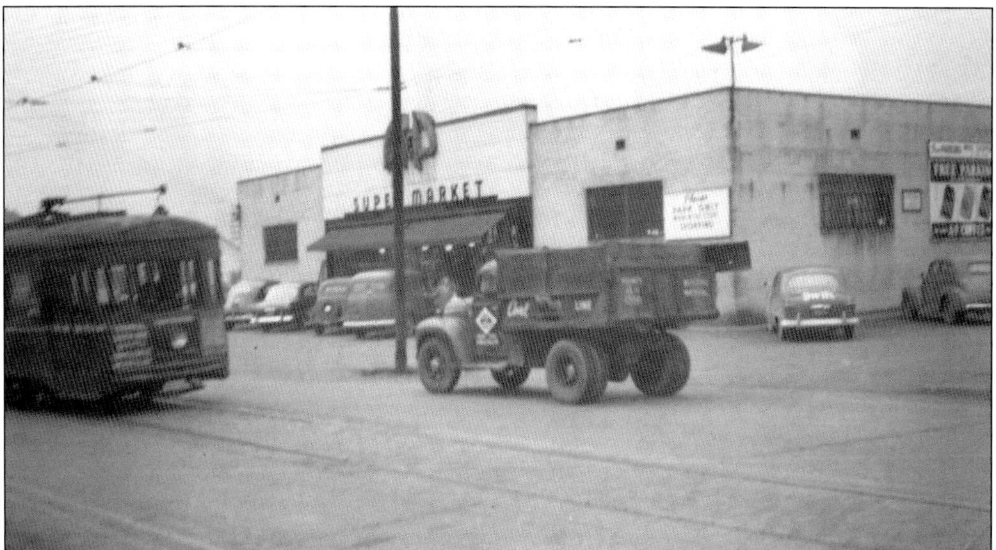

The first supermarket in Connellsville was this A&P store on South Arch Street near the West Penn terminal. In the 1950s, supermarkets began to gradually replace mom-and-pop stores, which had been part of the landscape in every Connellsville neighborhood. (Courtesy of Sandy Russell.)

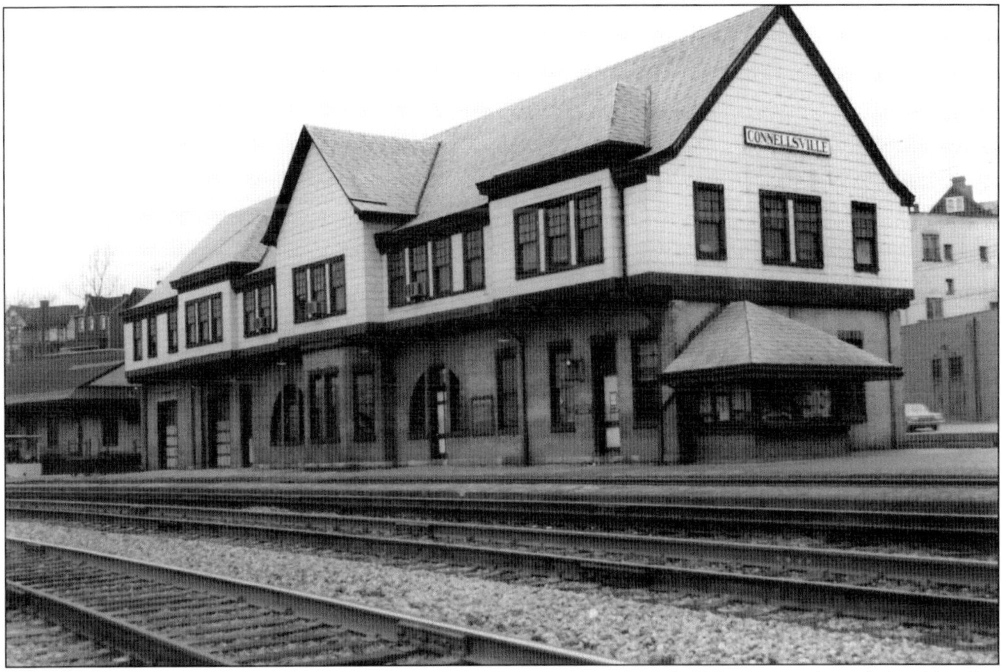

The B&O passenger station on Water Street is seen above just before it was demolished by the CSX Corporation in 1981. The original one-story structure was erected in 1897 and additional floors were added in 1904. The interior of the station (below) provided comfortable surroundings for passengers waiting to board famous B&O trains such as the *Capital Limited*, the *Ambassador*, the *Columbian*, and the *Shenandoah*.

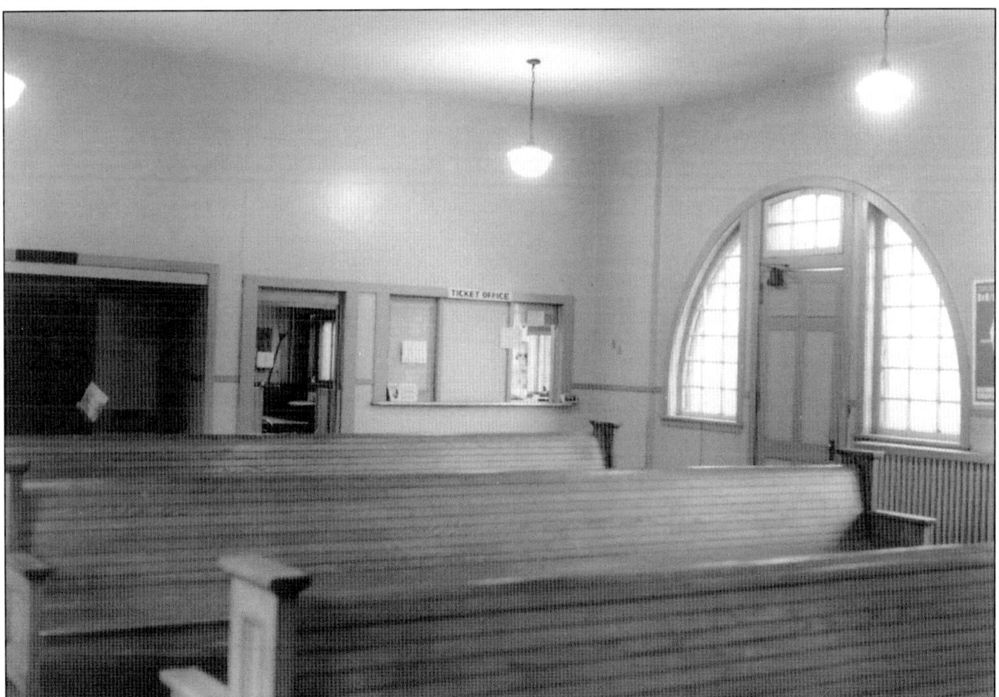

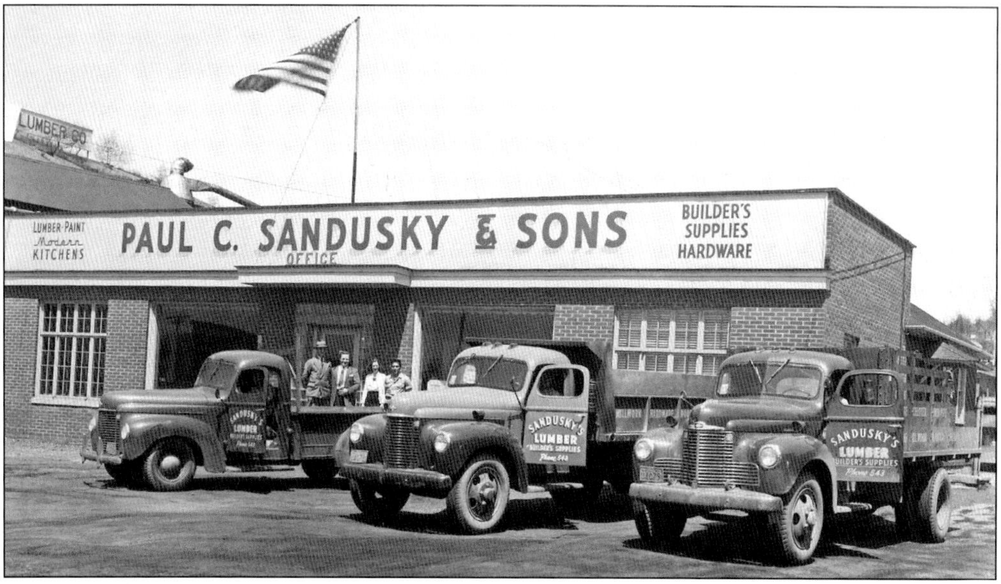

The Sandusky Lumber Company was a familiar landmark in the north end. Seen here standing in front of the office and showroom in 1948, are, from left to right, Paul Sandusky, Conrad Sandusky, Geneva Sandusky Riedman, and John Yakimick. The truck drivers are, from left to right, Robert Sandusky, Bernard Schlinger, and Joseph Stoots.

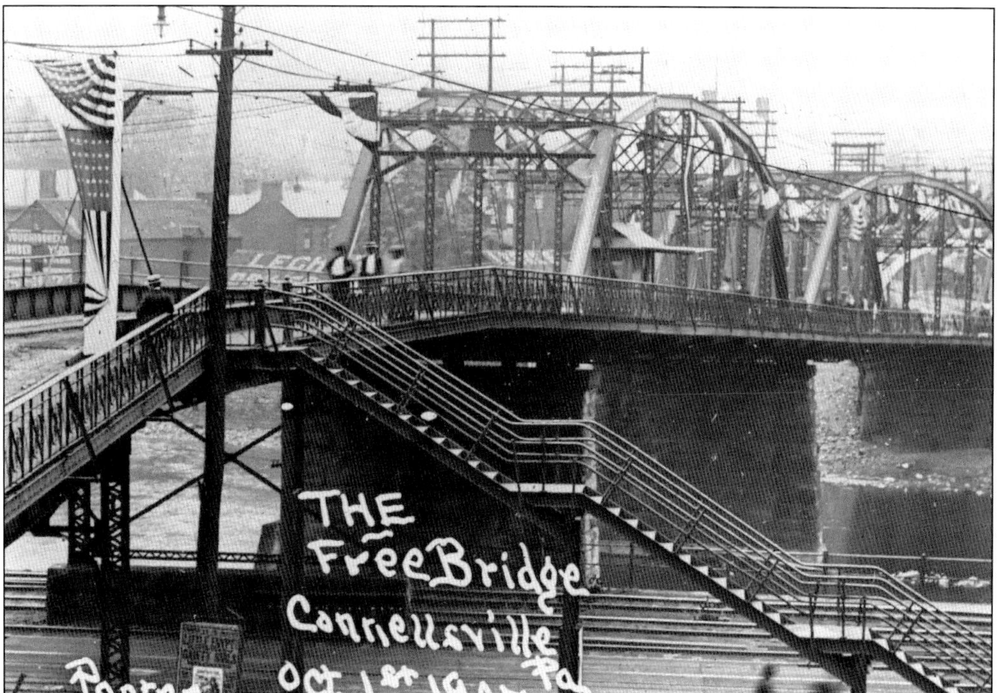

The "free bridge" connecting Connellsville with New Haven is seen in this 1907 H.C. Porter photograph. It was the fifth span across the Youghiogheny River at or near the beginning of West Crawford Avenue in Connellsville. The bridge was constructed in 1898 to replace the suspension bridge, which was limited in weight capacity and vehicular traffic.

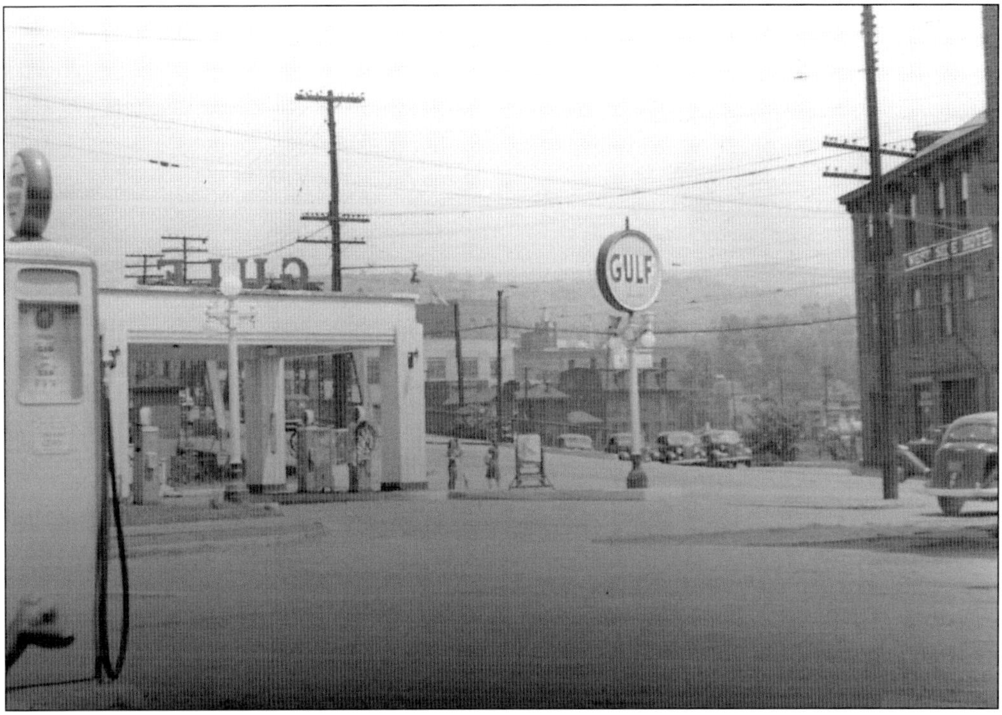

McAffee's Gulf station is seen here on the corner of First Street and West Crawford Avenue in the mid-1940s. The Westside Hotel is visible on the right. The hotel was later demolished and replaced by the offices of Widmer Engineering.

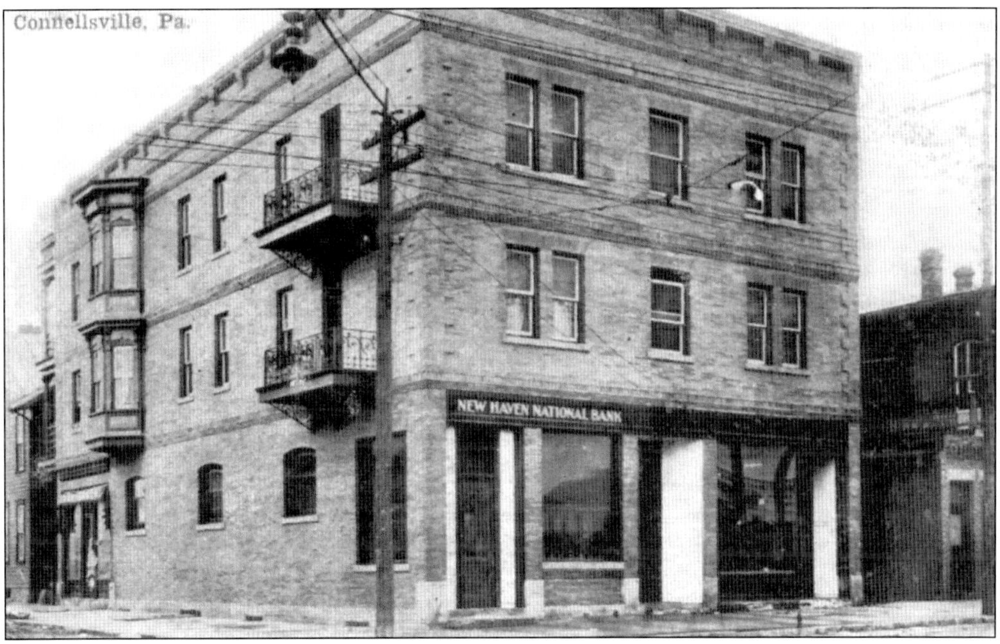

New Haven Bank was organized in 1902 and occupied this corner of West Crawford Avenue and Fourth Street. The bank was on the ground floor and apartments were on the upper levels. (Courtesy of Margaret Molinaro.)

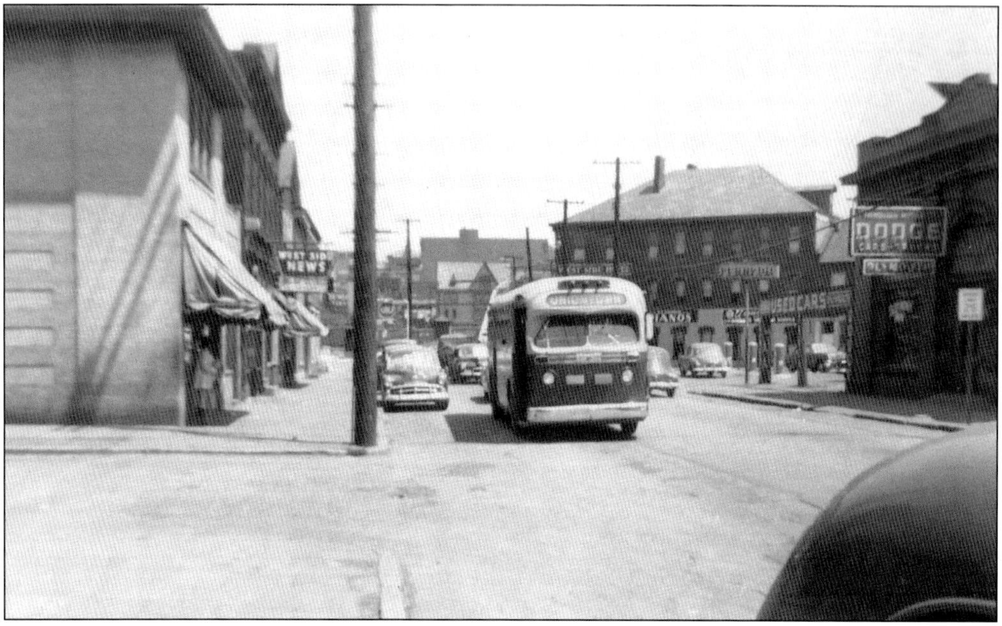

This photograph of West Crawford Avenue was taken in 1953 after buses replaced West Penn streetcars. On the left is a sign advertising the *Westside News*, which continues to serve neighborhood residents today. (Courtesy of Tillie Berg.)

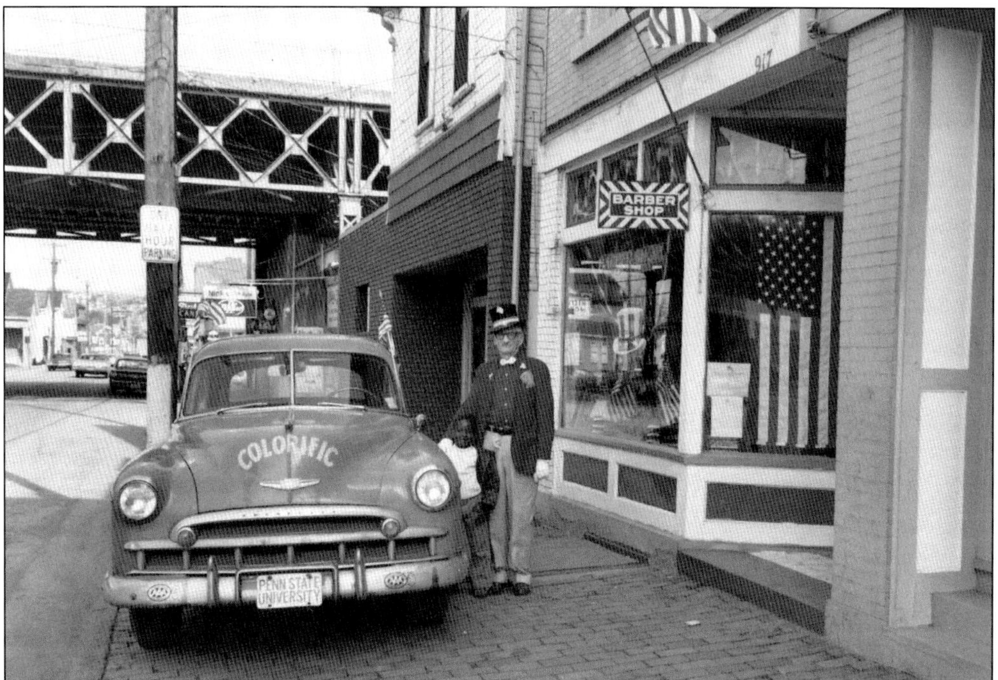

John Borchin stands between his car and his barbershop at 917 West Crawford Avenue in the early 1970s. Borchin moved to Connellsville from Vanderbilt and was noted for his unusual dress, which often featured patriotic themes. In his barbershop, customers could have their hair cut in a replica of an "electric chair." (Courtesy of Felix Prestia.)

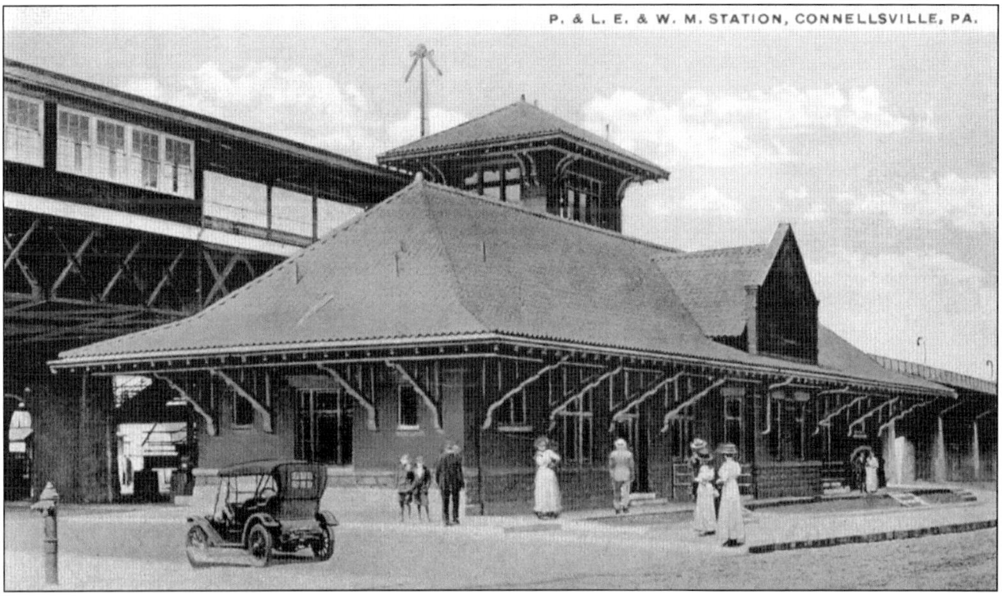

The Pittsburgh & Lake Erie Railroad (P&LE) passenger station and elevated platforms are seen here around 1920. The line to Connellsville was originally constructed by the Pittsburgh, McKeesport & Youghiogheny Railroad and subsequently leased to the P&LE in 1884. In 1912, the Western Maryland Railway laid tracks from Cumberland, Maryland, to the Connellsville station, thereby completing a vital link for the P&LE between Pittsburgh and Baltimore.

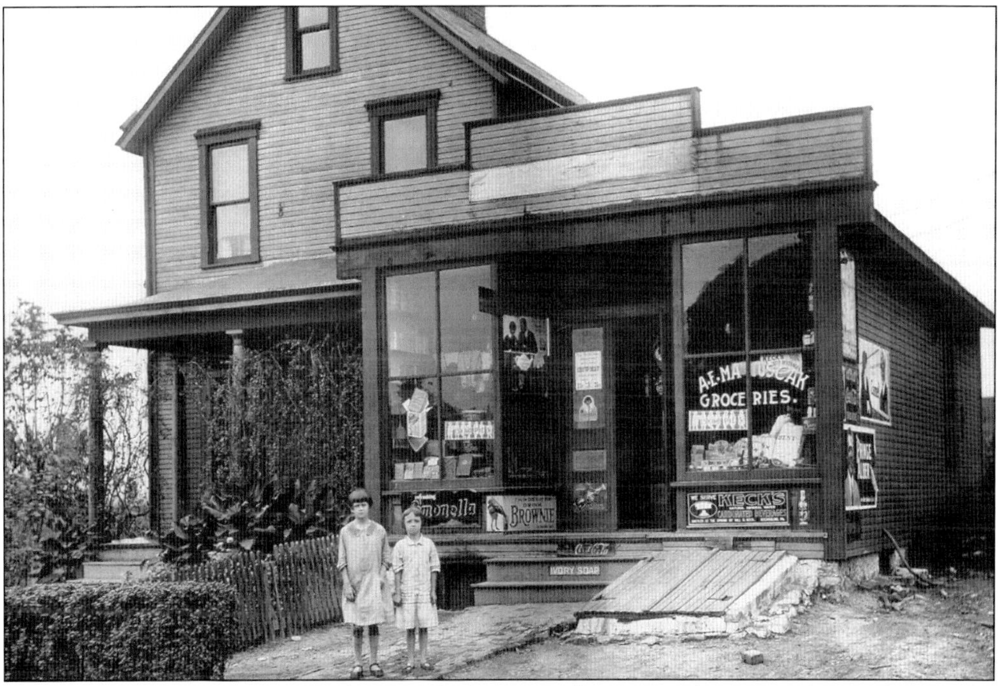

Margaret (left) and Frances (right) Matuschak are seen here around 1915 standing in front of the grocery store operated by their parents at 1520 West Crawford Avenue, in the Brookvale section of Dunbar Township. The store opened in 1908 and closed in 1975. It was typical of the era when most neighborhoods had family-owned corner businesses. (Courtesy of Carmine Molinaro Jr.)

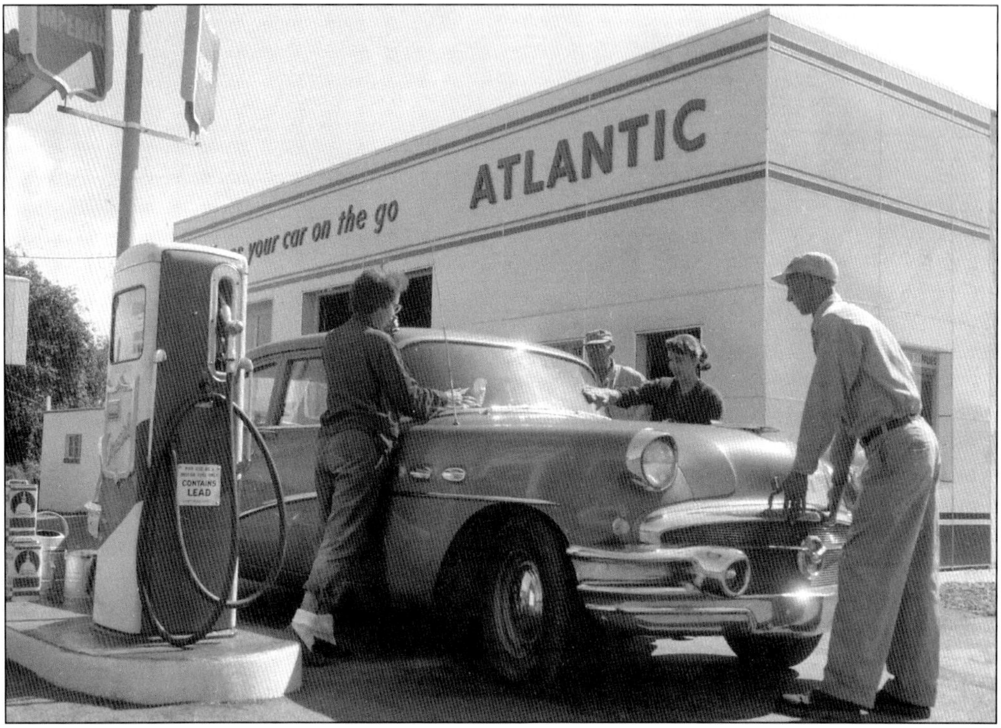

William Husband (right) operated this gas station between Connellsville and Uniontown on Route 119 in 1955. The station "attendants" are Edna Husband (left) and Janet Husband. Ben Husband, who opened the first station on this site in 1944, is standing next to Janet. (Courtesy of John Husband.)

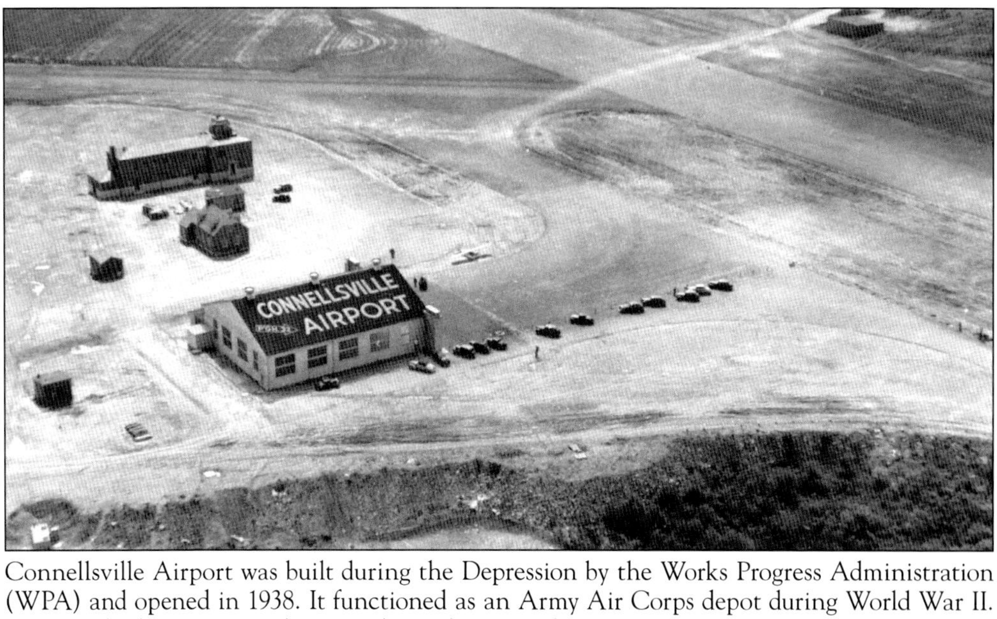

Connellsville Airport was built during the Depression by the Works Progress Administration (WPA) and opened in 1938. It functioned as an Army Air Corps depot during World War II. Afterwards, the airport was home to the Taylor Aircraft Corporation, a US Marine Corps training center, and other enterprises. In 1965, Fayette County assumed management of the facility. In 2007, the name was changed to the Joseph A. Hardy Connellsville Airport.

Three

Serving at Home and Abroad

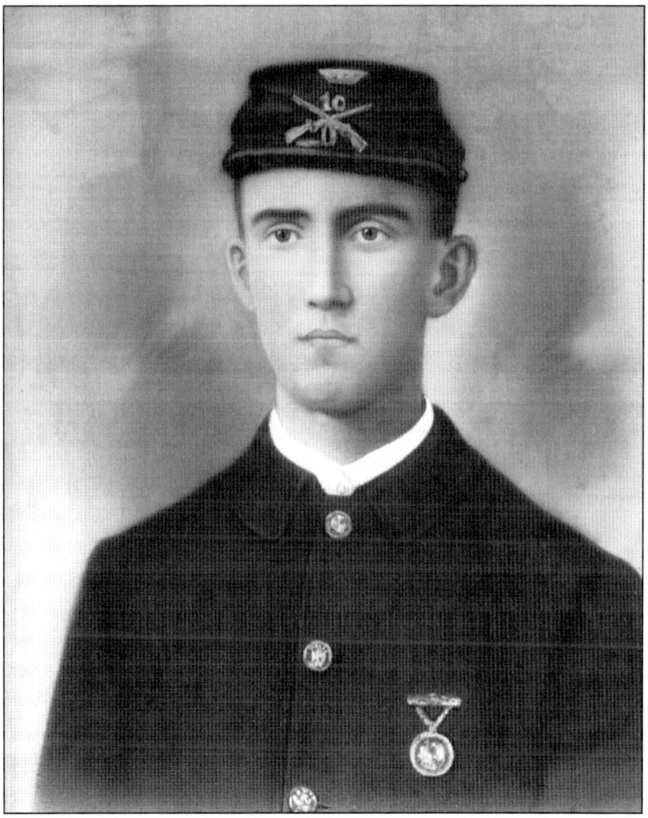

Walter Brown was the first Connellsville resident killed in a foreign war. He enlisted in Company D, 10th Regiment, Pennsylvania Volunteers, on April 27, 1898, during the Spanish-American War. Brown lost his life on July 31, 1898, during the Battle of Malata in the Philippines. Veterans of Foreign Wars Post 21 is named in his honor. (Courtesy of Veterans of Foreign Wars Post 21.)

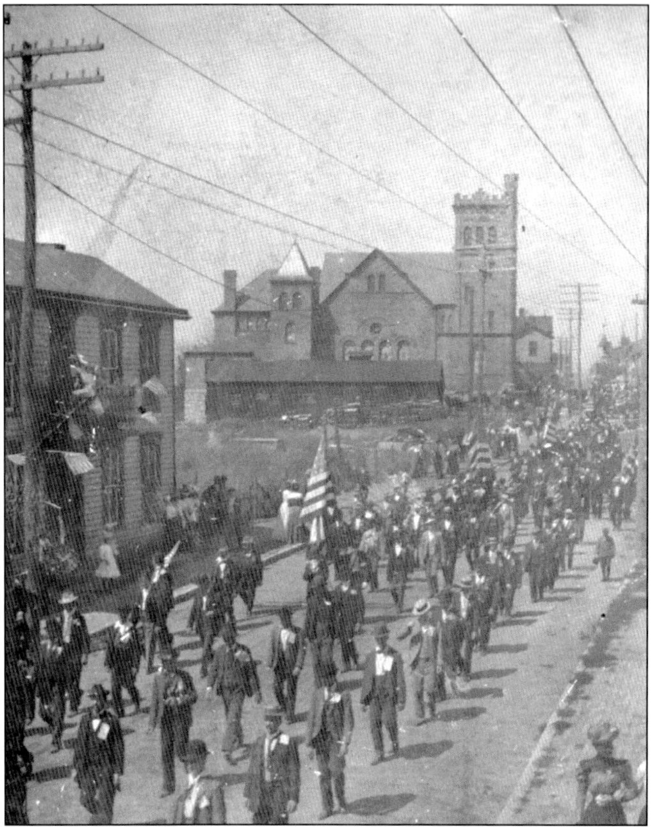

Civil War veterans and others parade on South Pittsburgh Street in the 1890s. During the war, 16 companies were raised from Fayette County, including Company H, from Connellsville. The company was assigned to the 142nd Regiment of the Pennsylvania Volunteer Infantry. The regiment was deployed at Fredericksburg, Chancellorsville, Gettysburg, Wilderness, and Petersburg.

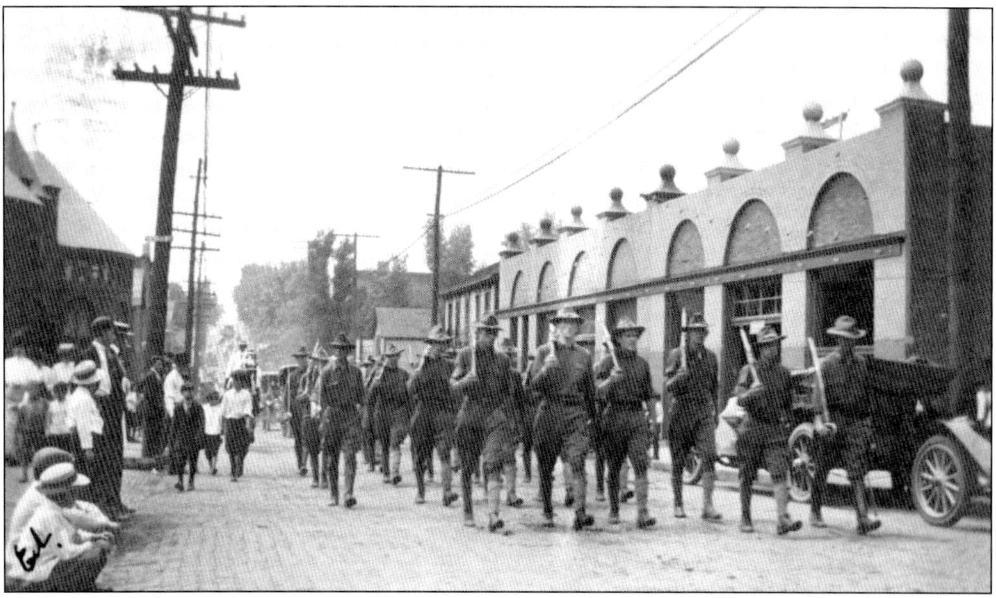

Connellsville residents lined South Pittsburgh Street to celebrate the end of the war in 1919 and to honor veterans from around Connellsville. The Baptist church is on the left. The Yough Ford Motor Company is on the right. There are stories about Henry Ford and Thomas Edison stopping at the company in order to have their vehicles serviced during hunting trips in the region.

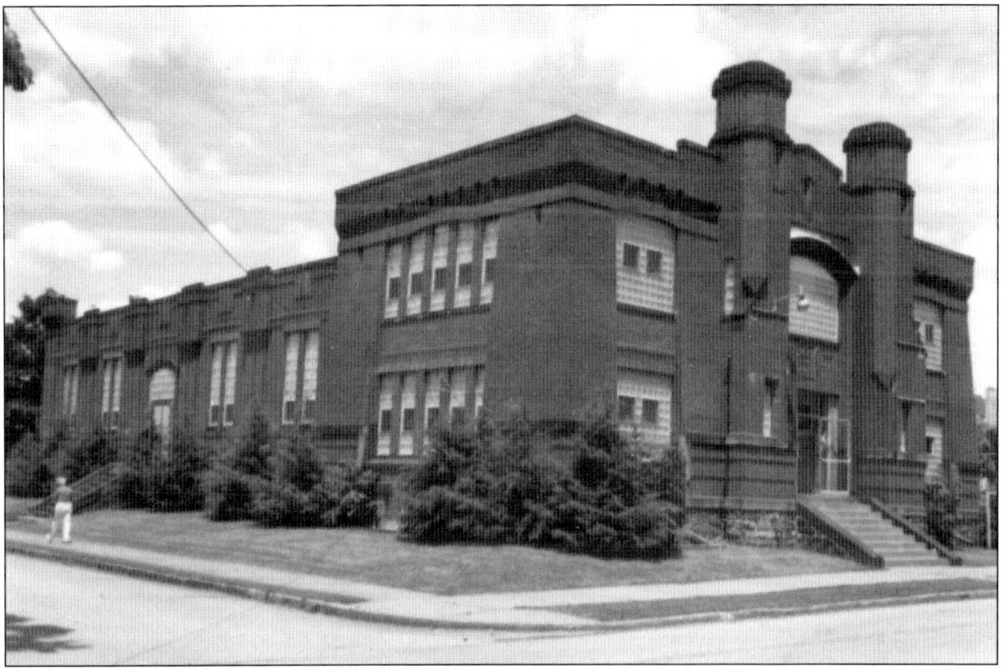

The Connellsville armory, located on West Washington Avenue, dates from 1907. Initially, it was the headquarters for Company D, 10th Infantry Regiment. During World War I, it was renamed the 110th Infantry Regiment, Pennsylvania National Guard. The facility closed when the National Guard moved to the Readiness Center in 2005. In addition to hosting military activities, the building also served as a venue for many community events, including Immaculate Conception High School basketball games.

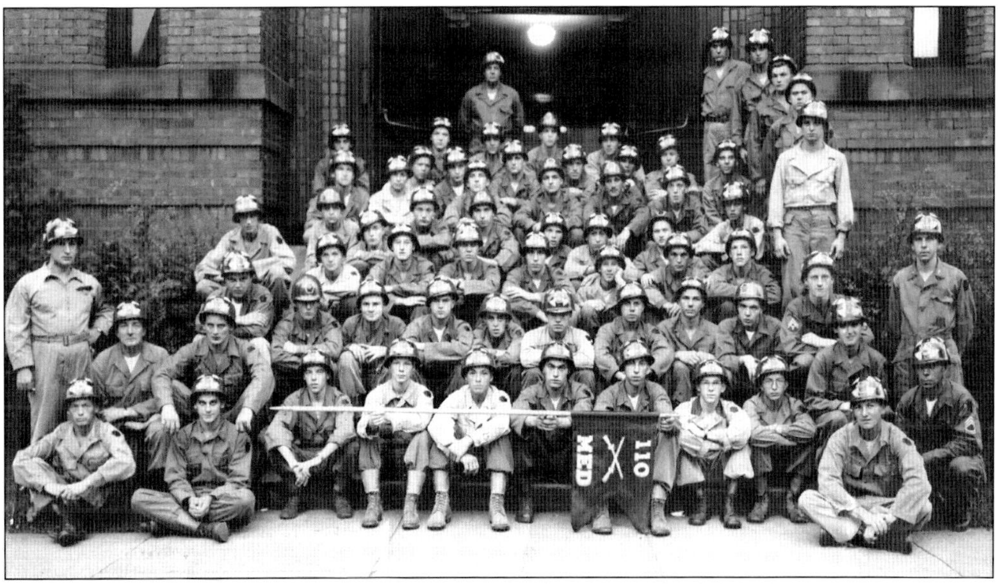

The Connellsville armory was an operational base for this medical detachment in 1949, which was part of the 28th, or Keystone, Division. The 28th Division originated during the Civil War and was engaged in the Spanish-American War and two world wars. In World War II, it rendered distinguished service at Normandy, the Hurtgen Forest, and the Battle of the Bulge.

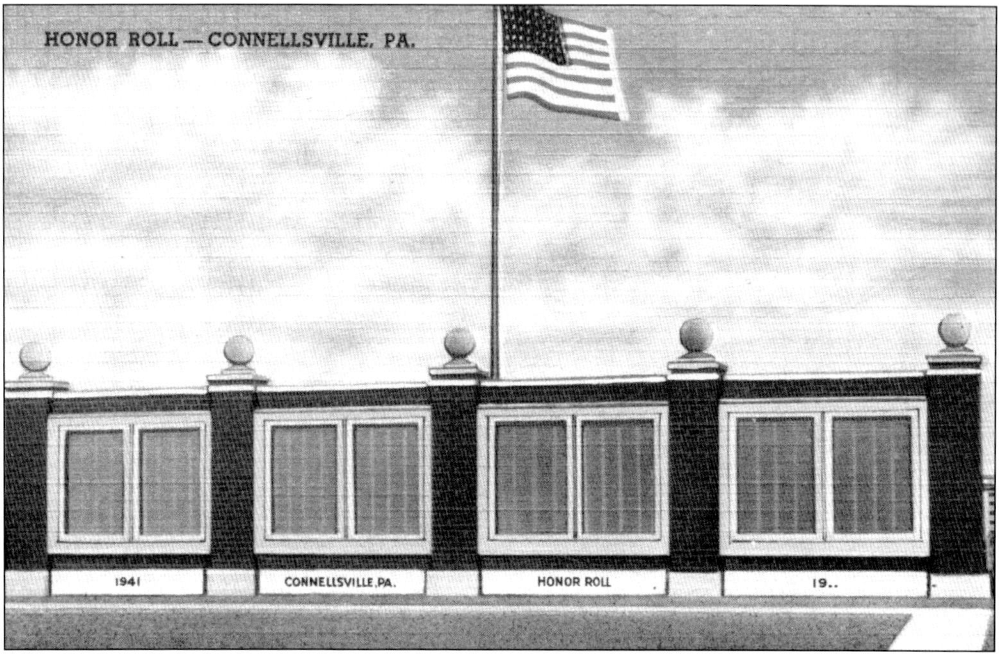

This postcard shows the honor roll erected by Connellsville residents in memory of veterans who had served in foreign wars. It was located on the property that is now Lion's Square.

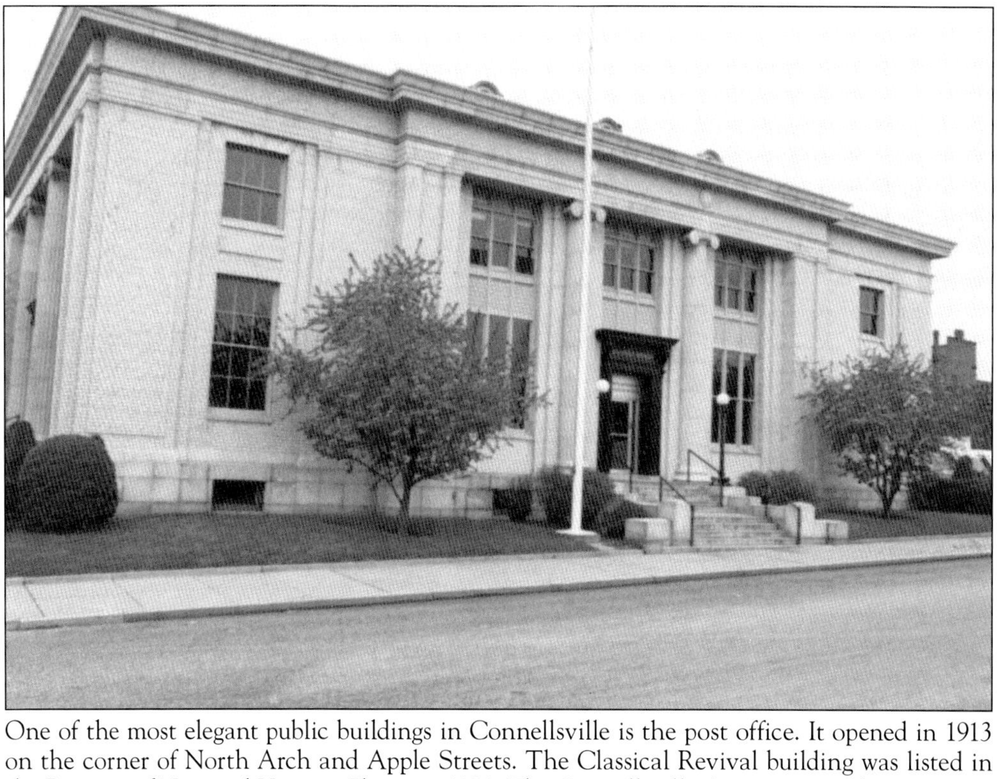

One of the most elegant public buildings in Connellsville is the post office. It opened in 1913 on the corner of North Arch and Apple Streets. The Classical Revival building was listed in the Register of National Historic Places in 1994. The Connellsville Area Historical Society and postmaster C. Franklin Ellis led the effort to secure this recognition.

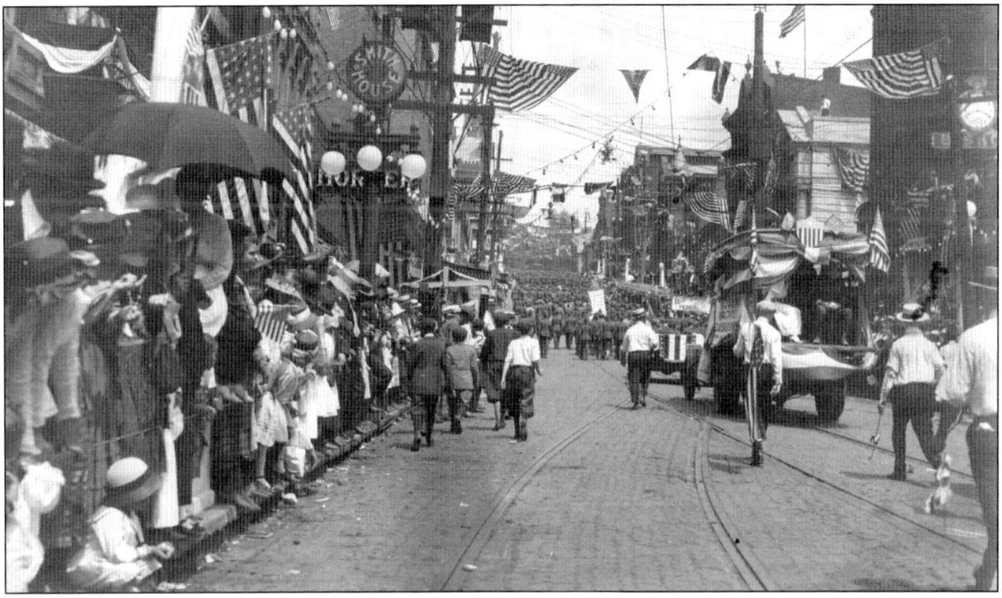

Veterans march up the hill on East Crawford Avenue in 1919, celebrating the end of World War I.

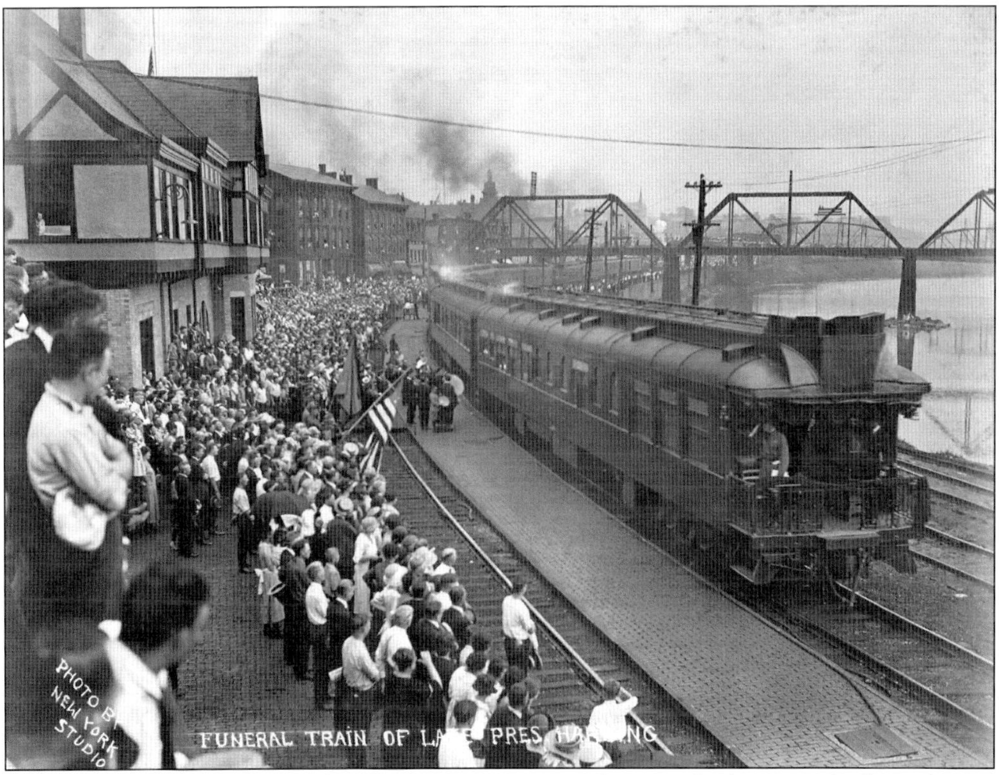

Gen. John J. Pershing is seen here in August 1923 standing on the last car of a funeral train that carried the body of Pres. Warren J. Harding from San Francisco to Washington, DC. He is addressing a huge crowd gathered at the B&O passenger depot on Water Street in a show of respect for the president. (Courtesy of Joe Kurtz.)

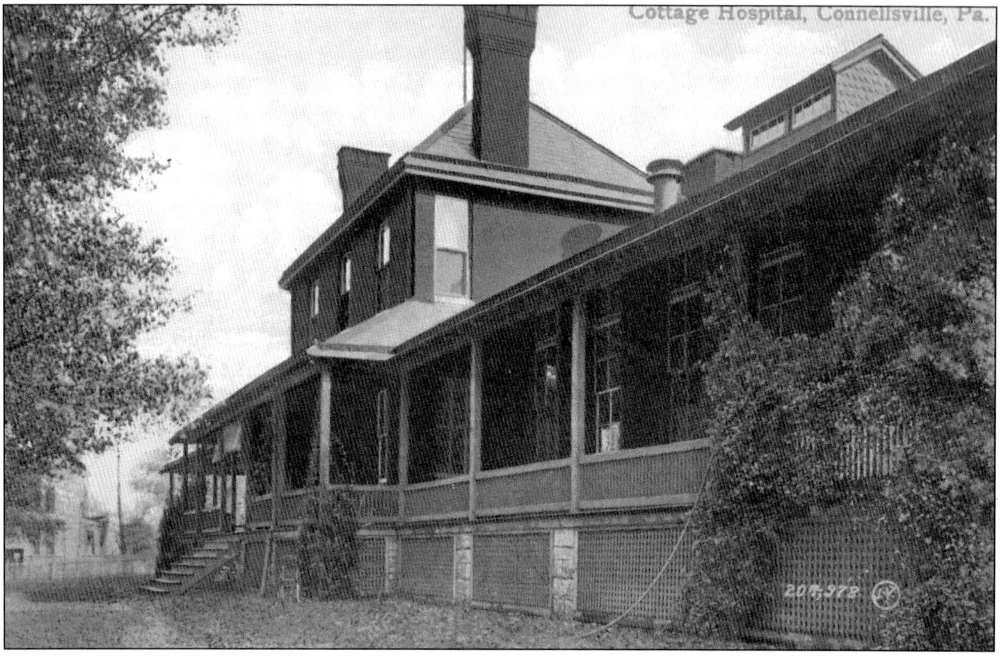

The Cottage State Hospital (above) opened on East Murphy Avenue in 1891 with 38 beds. Connellsville was one of four cities in Pennsylvania to receive funding for a hospital in 1887, pursuant to an act of the Pennsylvania General Assembly. Initially, the facility occupied two acres on high ground above the congestion of the central business district. Nurses and a patient are seen below on front steps of the institution in 1906. In the beginning, most patients were coal miners and railroad workers who had been injured on the job. Companies paid $3.50 per week for treatment. (Above, courtesy of Margaret Molinaro; below, courtesy of Barbara and Tony Keefer.)

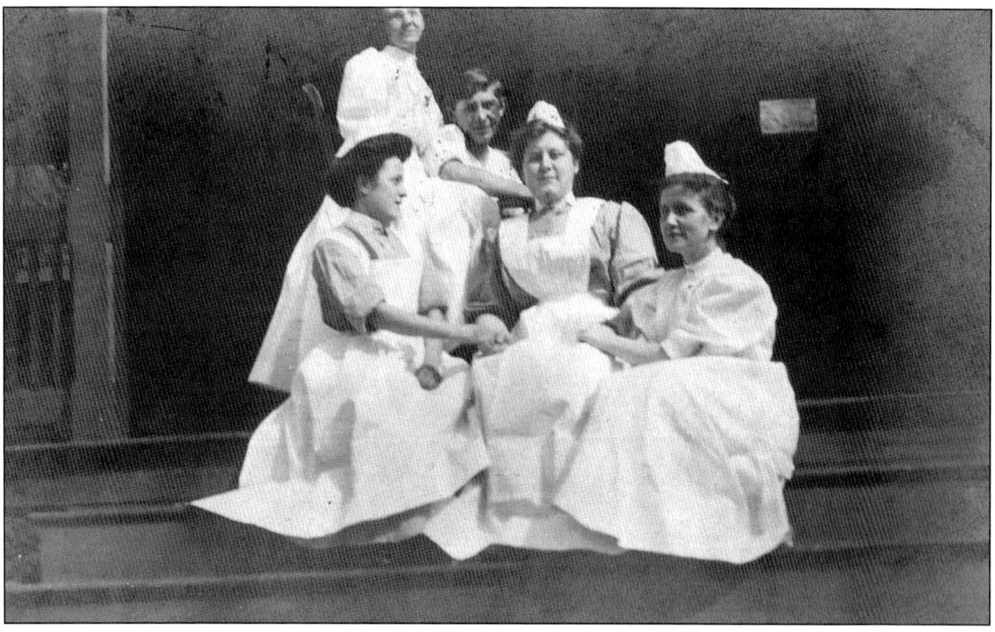

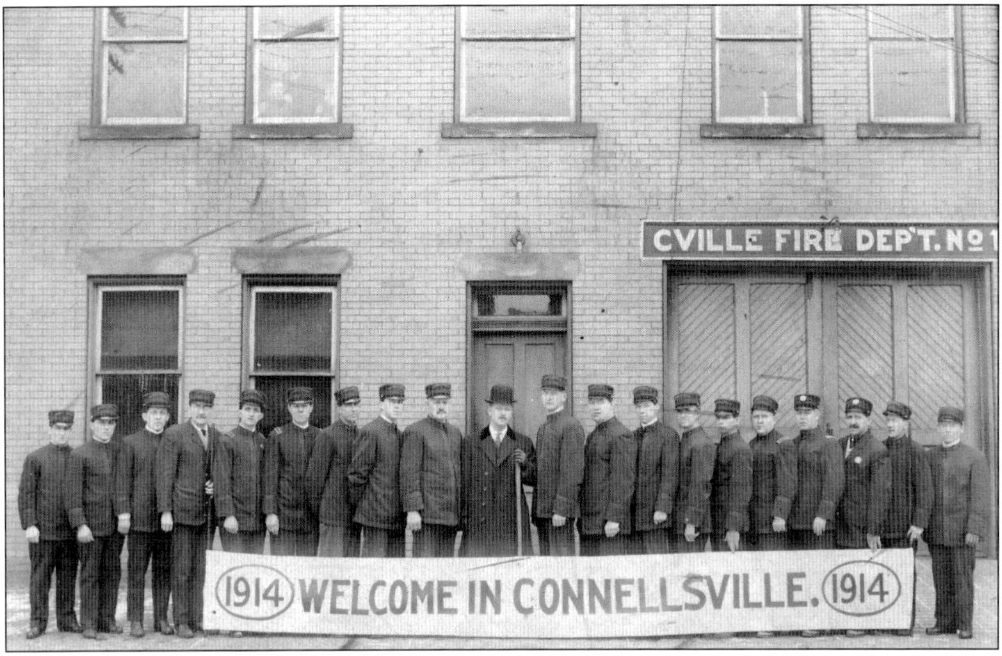

Members of the Connellsville Fire Department assemble in front of the west side station in 1914 with a welcome sign for visiting companies. The department was organized in 1811. By the time of the centennial, there was a company in each of the four wards, consisting of both paid employees and volunteers.

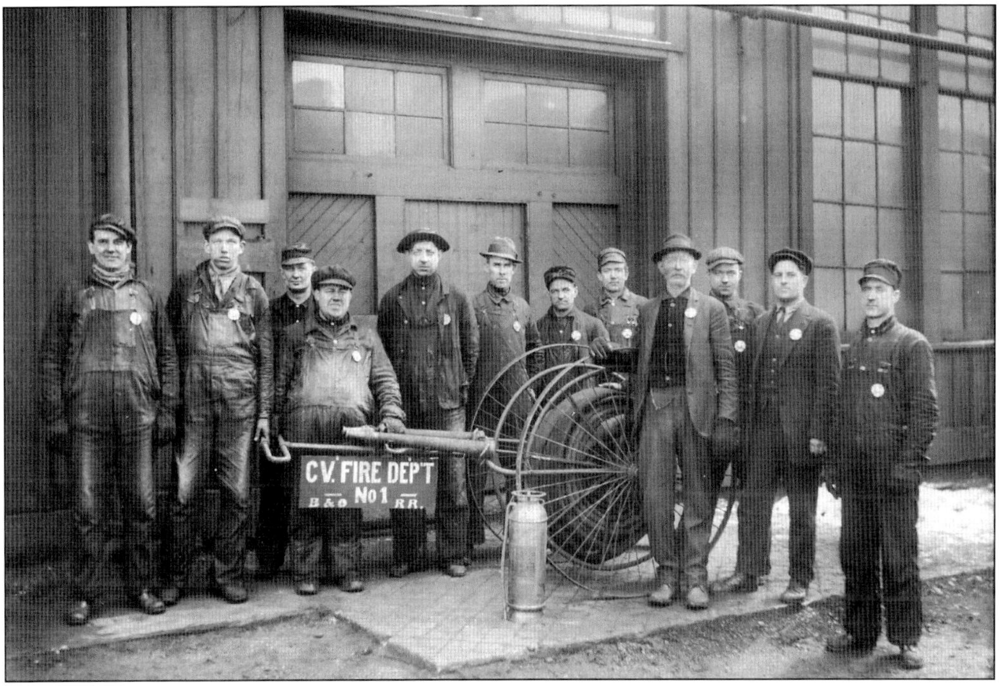

Fire was a constant danger in and around the B&O roundhouse in Connellsville. Consequently, the railroad maintained its own fire department, which is seen here in 1901. James E. Murphy is the fifth department member from the left. The others are unidentified.

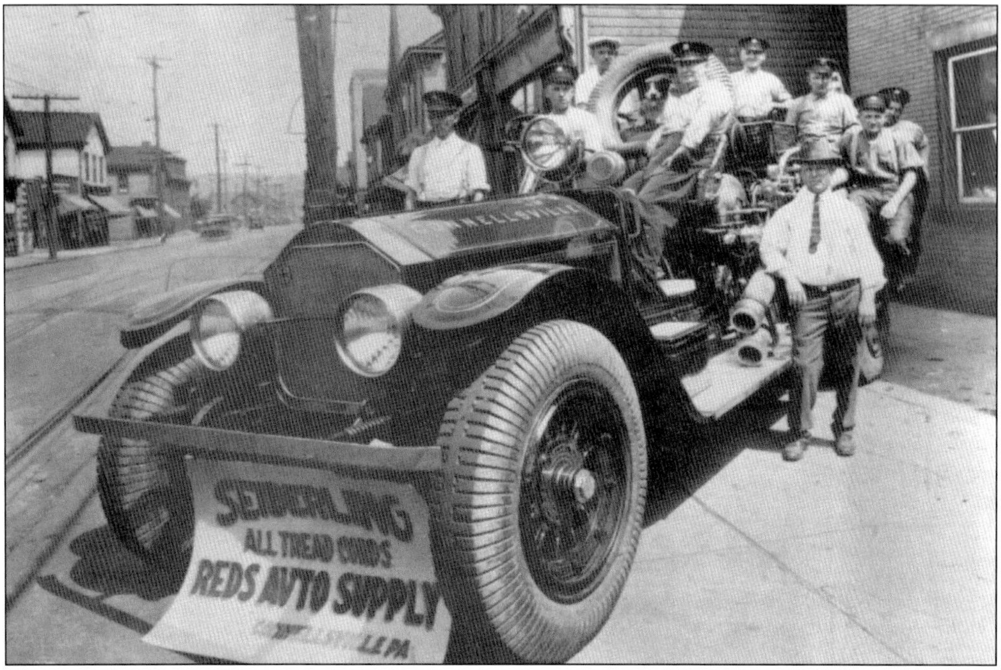

Connellsville Fire Department employees and volunteers display their third motorized truck in 1924 on West Crawford Avenue. They are, from left to right, (first row) W.E. DuBolt (chief), John "Buck" Giffin (driver), Harry Cypher, and Malaky Duggan; (second row) John Giffin, Tom Dailey, Harry Bower, George Shumaker, and J.M. Martin. The sign in front of the engine advertises "Red" Sturbutzel's store, which supplied the tires. (Courtesy of New Haven Hose Company.)

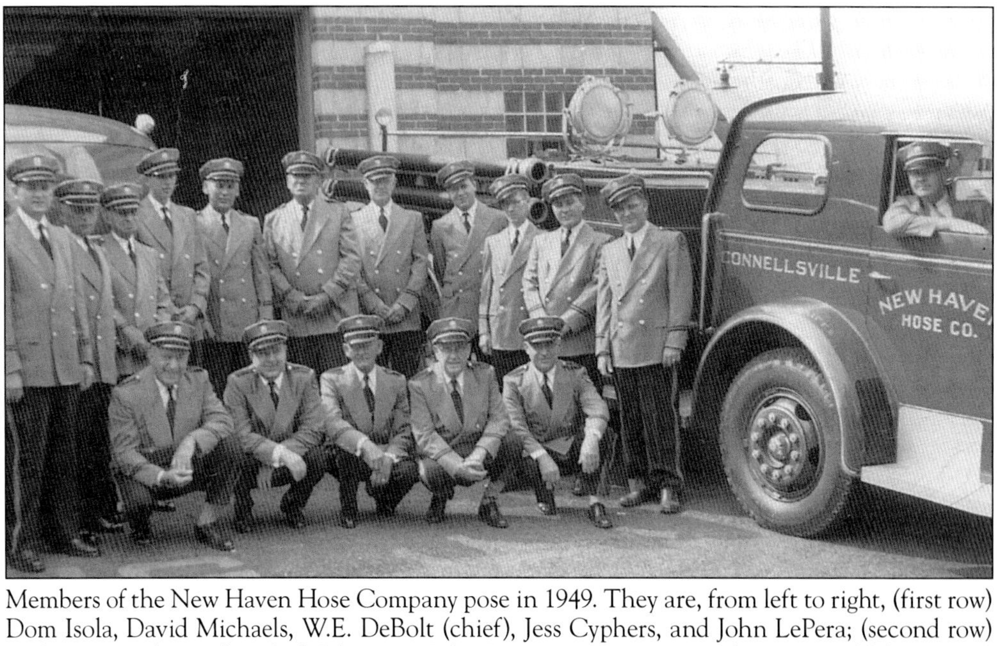

Members of the New Haven Hose Company pose in 1949. They are, from left to right, (first row) Dom Isola, David Michaels, W.E. DeBolt (chief), Jess Cyphers, and John LePera; (second row) Melvin De Bolt, unidentified, John Nez, John Dunston, Martin Fronek, Harry Schlebey, Frank Gaiduser, James McNulty, Harry De Bolt, John Fiano, and Frank O'Brien. Ralph Strickler is sitting in the truck. (Courtesy of New Haven Hose Company.)

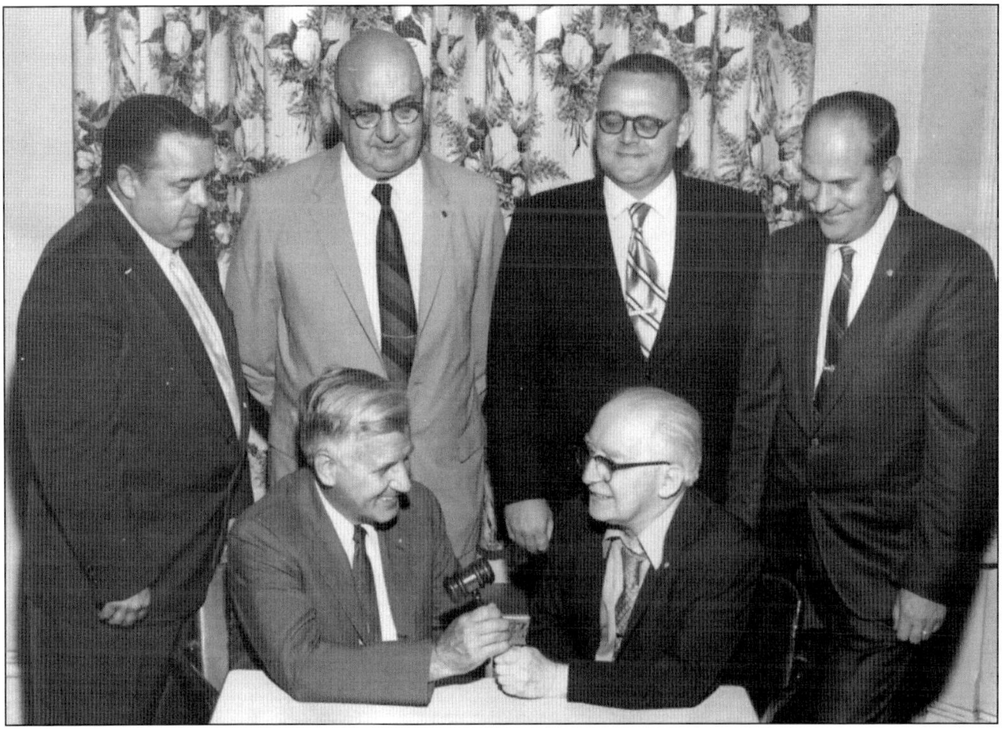

Above, Max Floto (sitting at right) presents a gavel to Don Cope, the newly elected president of the Connellsville Veterans Commission in 1972. Other veterans observing the transition are, from left to right, Harry Jones, Ralph Burkett, Jack Kopf, and Clarence Smith. Floto was a World War I veteran who, along with Thomas Scott Sr., lobbied for the establishment of Armistice Day (now Veterans Day) as a national holiday. (Courtesy of Ed Cope.)

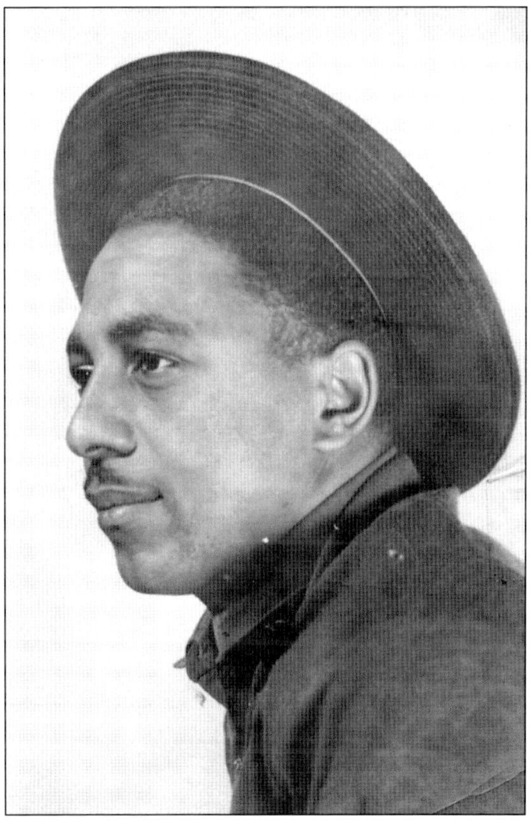

Sylvester Parrish of Dunbar was a Tuskegee Airman, a group of some of the first African Americans to serve as pilots or in ground support roles in the US Army Air Corps during World War II. The Tuskegee Airmen trained at the Tuskegee Institute in Alabama. After the war, Parrish operated an accounting business and was a trustee of the Payne African Methodist Episcopal Church in Connellsville. (Courtesy of Ted Davis.)

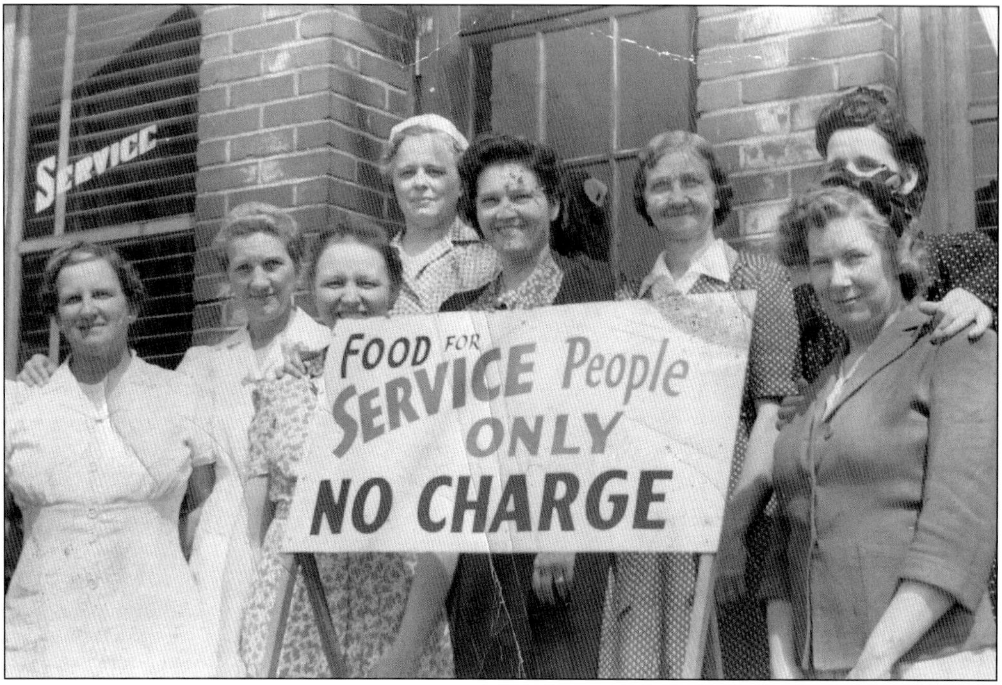

Seen above are many of the women who operated the famous Connellsville Canteen, located on Water Street across from the B&O Railroad passenger station. Between 1944 and 1946, women from Connellsville volunteered to prepare and serve food to all military personnel traveling through the city on troop trains. They used their rations and donations from grocery stores and farms to feed nearly 500,000 men and women in uniform during World War II. Below, the "canteen ladies," who rendered an invaluable service on the home front as part of the war effort, posed for a group photograph in 1945. Ruth Kunkle, a section leader in the organization, is the eighth person from the left in the second row. The others are unidentified.

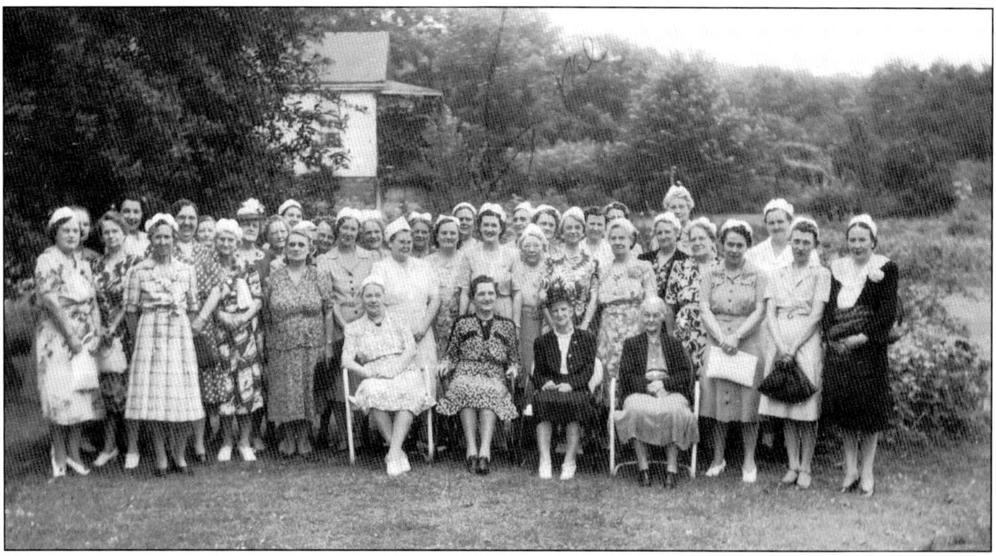

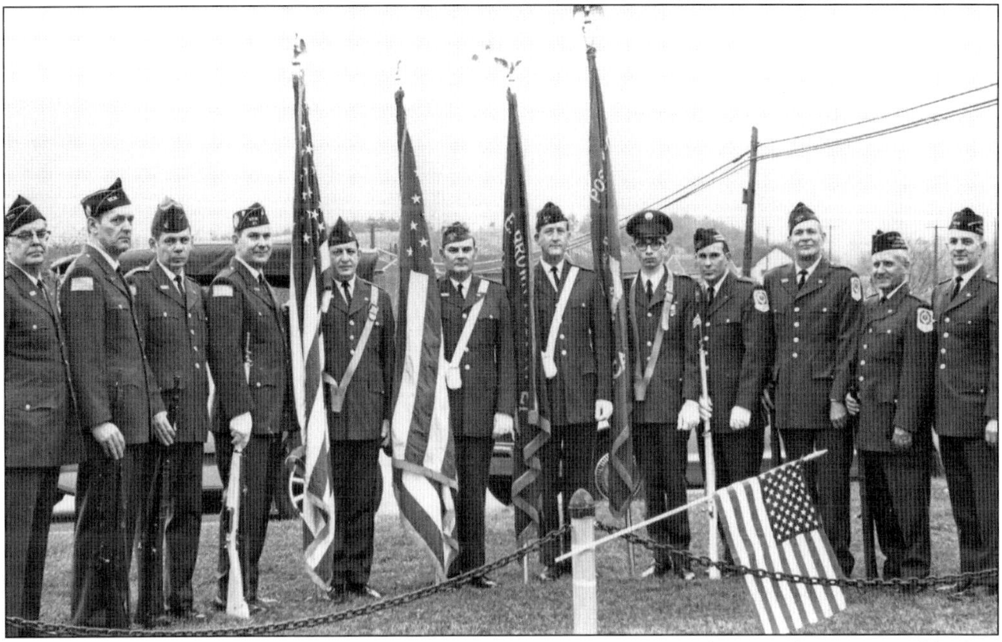

Members of the Connellsville Veterans of Foreign Wars Post 21 form an honor guard before Memorial Day ceremonies in the late 1950s. They are, from left to right, Harry Trevitt, John Smith, James Hickey, Clarence Smith, Ray Huey, Albert Furtney, Fred Luckey, unidentified, Don Smith, James Smith, Don Cope, and Allen Jones. (Courtesy of Ed Cope.)

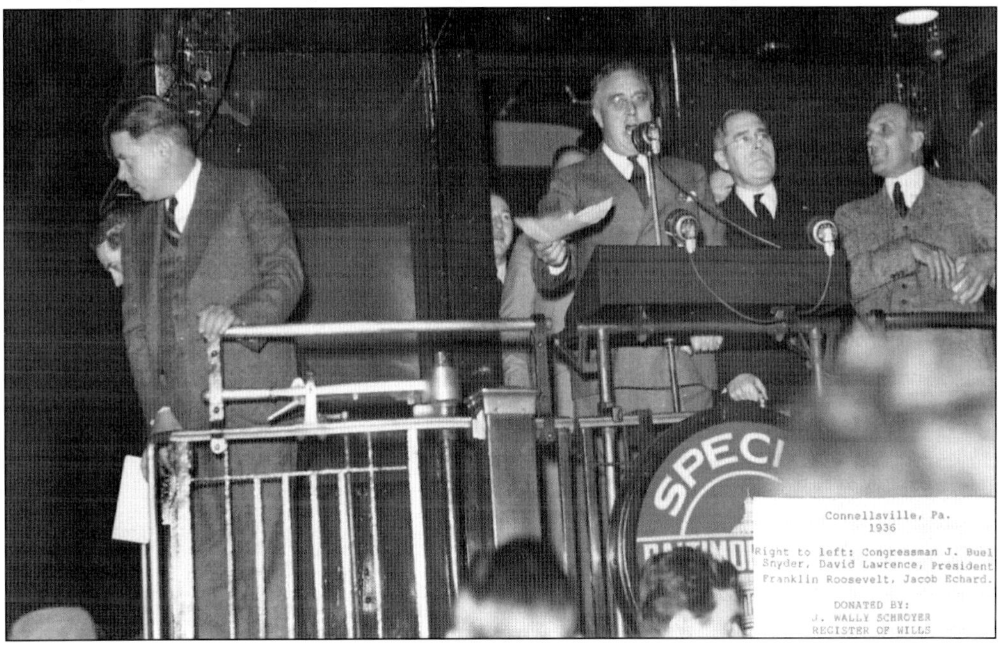

On his way to his landslide reelection in 1936, Pres. Franklin D. Roosevelt made a campaign stop at the B&O station in Connellsville. Standing to the right of the president are David Lawrence and Congressman J. Buell Snyder, who represented the 24th District. At the time, Lawrence was the chairman of the Pennsylvania Democratic Party. He went on to serve as the mayor of Pittsburgh and the governor of Pennsylvania.

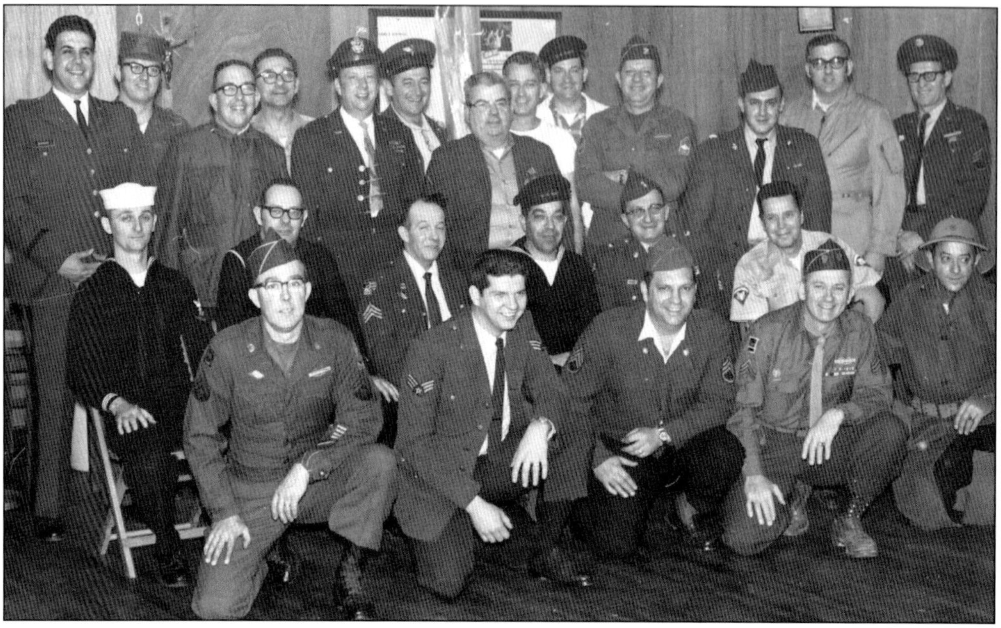

Knights of Columbus members wore their military uniforms to this 1965 meeting. They are, from left to right, (first row) Bud King, Dwire Sheen, Pete Hartz, Bob McGinnis, and unidentified; (second row) Joe Zavatchen, Walt Winters, Joe Trainer, George Pilla, ? Caringolo, and Blane Reed; (third row) ? Canada, Andy Hornick, Don Hefernan, Tony Ruse, Anthony Sikora, Gene Galiardi, Pat Bradley, Richard McGinnis, unidentified, John Fiesta, Amedio Masciarelli, and two unidentified men. (Courtesy of Felix Prestia.)

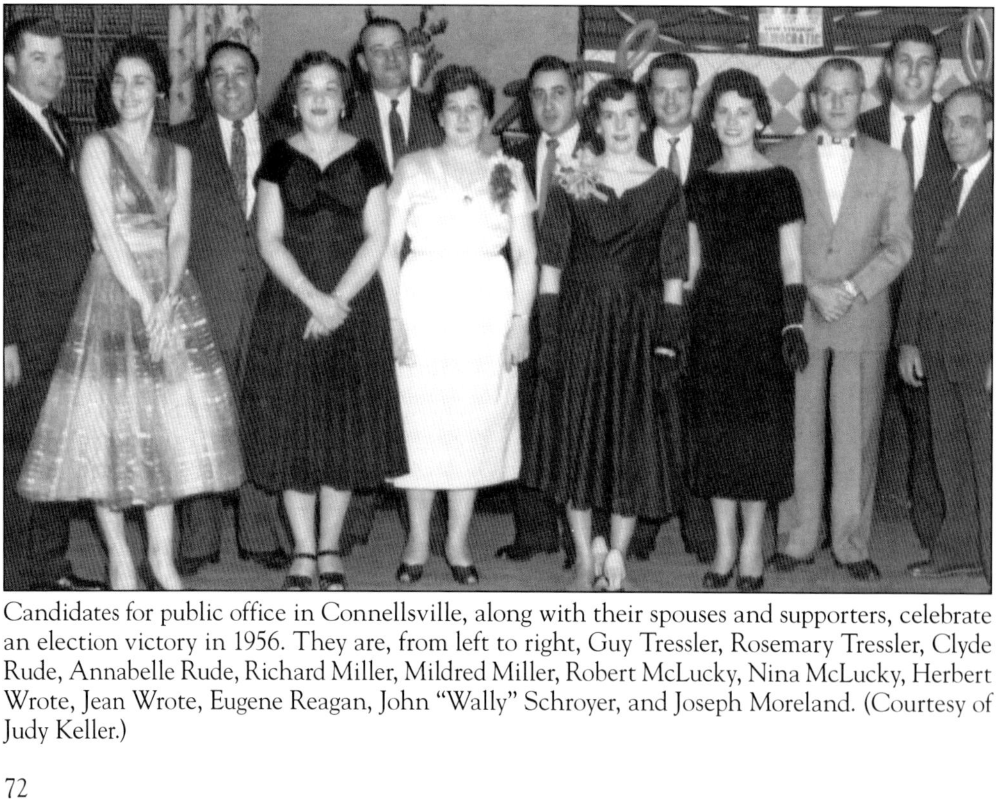

Candidates for public office in Connellsville, along with their spouses and supporters, celebrate an election victory in 1956. They are, from left to right, Guy Tressler, Rosemary Tressler, Clyde Rude, Annabelle Rude, Richard Miller, Mildred Miller, Robert McLucky, Nina McLucky, Herbert Wrote, Jean Wrote, Eugene Reagan, John "Wally" Schroyer, and Joseph Moreland. (Courtesy of Judy Keller.)

Fayette County commissioner Carmine Molinaro Jr. (right) meets with longtime police chief Anthony Cataldi (center) and Connellsville mayor J. Harold Dull to discuss training for the Connellsville Police Department in 1970. Molinaro was the youngest person elected county commissioner and served in the position from 1972 to 1987. (Courtesy of Carmine Molinaro Jr.)

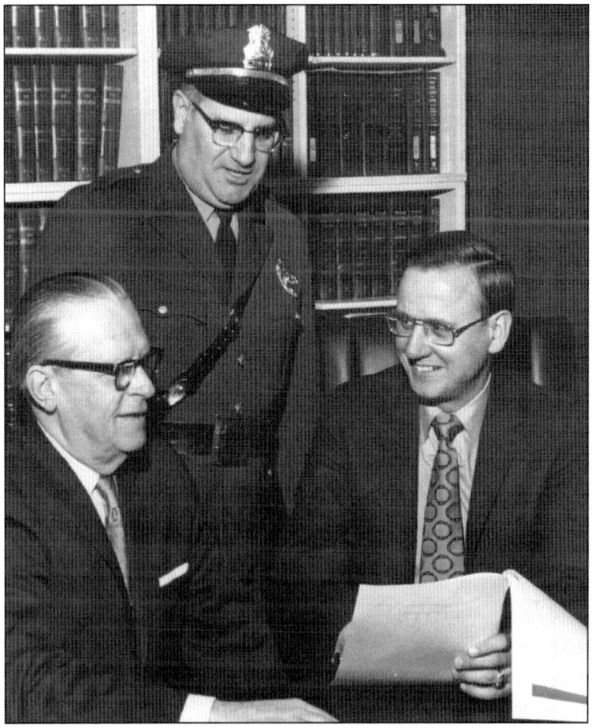

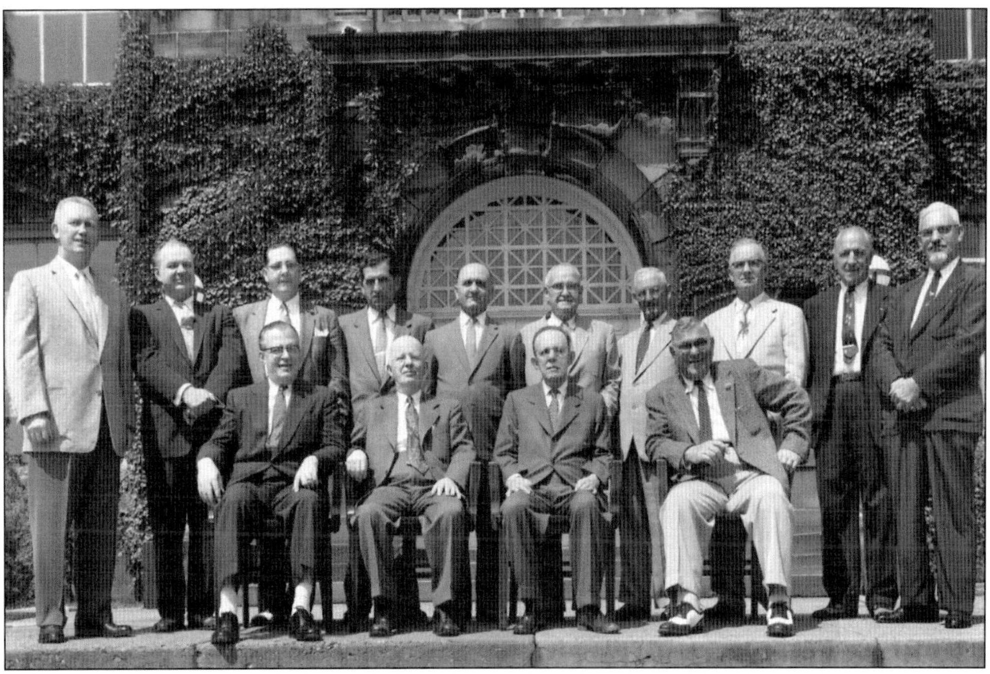

The Connellsville sesquicentennial committee poses in front of the high school on East Fairview Avenue in 1956. They are, from left to right, (seated) Dr. J. Harold Dull, Judge Ross Matthews, Meyer Aaron, and Connellsville mayor Abe Daniels; (standing) Dr. Ned Culler, D.W. Michaels, Milton Munk, Vincent Cuneo, Louis DeAuria, Jesse Cypher, William DeBolt, Dr. Daniel Minerd, Ernest Kooser, and I.D. Younkin. James Driscoll is missing from the photograph.

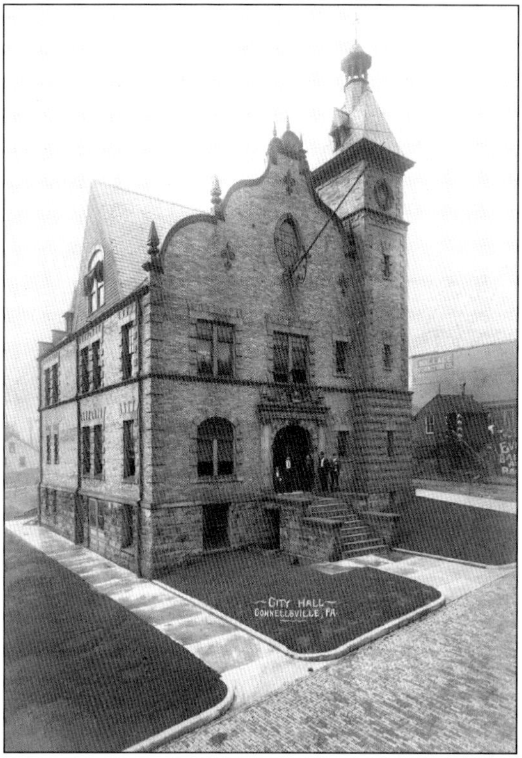

The first city hall was on the corner of Crawford Avenue and Pittsburgh Street. Larger accommodations were constructed in 1858 on Pittsburgh Street; however, citizens objected to housing prisoners in one of the most public sections of the city. In response, the council awarded a contract in 1901 to erect this new municipal office building. It was built in the Dutch architectural style on property originally donated by Zachariah Connell for public use.

Below, Edward Shaw (third row) poses with members of the Connellsville Police Department around 1936. Shaw was employed as a janitor at the department. The officers are unidentified. (Courtesy of John Ed Dailey.)

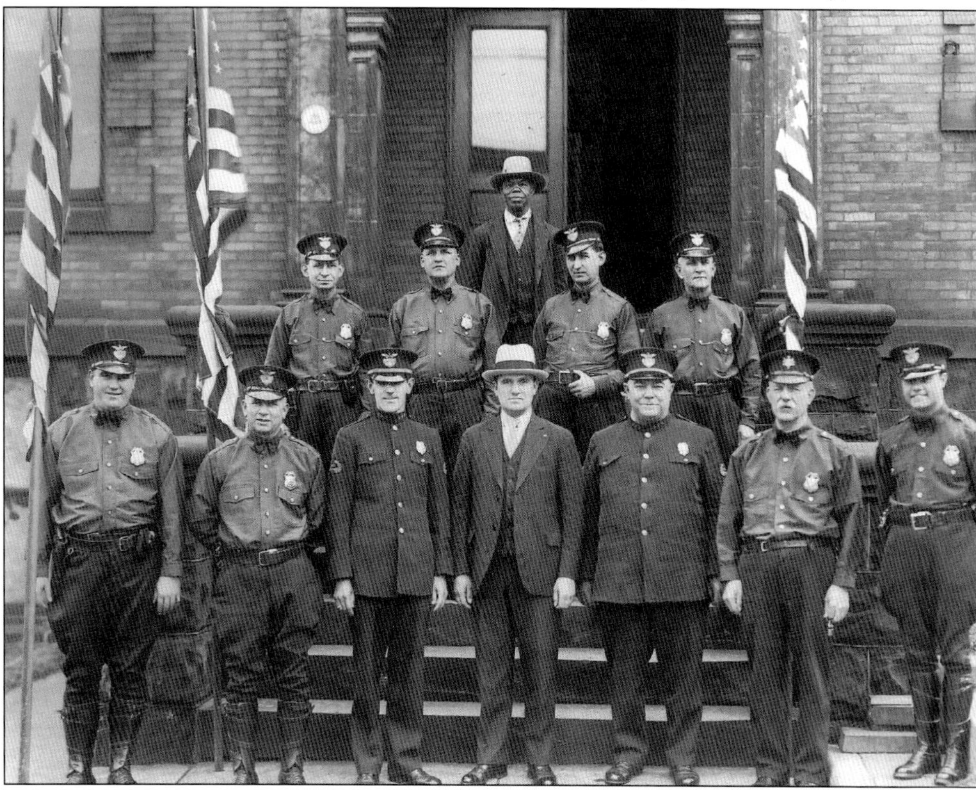

Four

LEARNING AND BELONGING

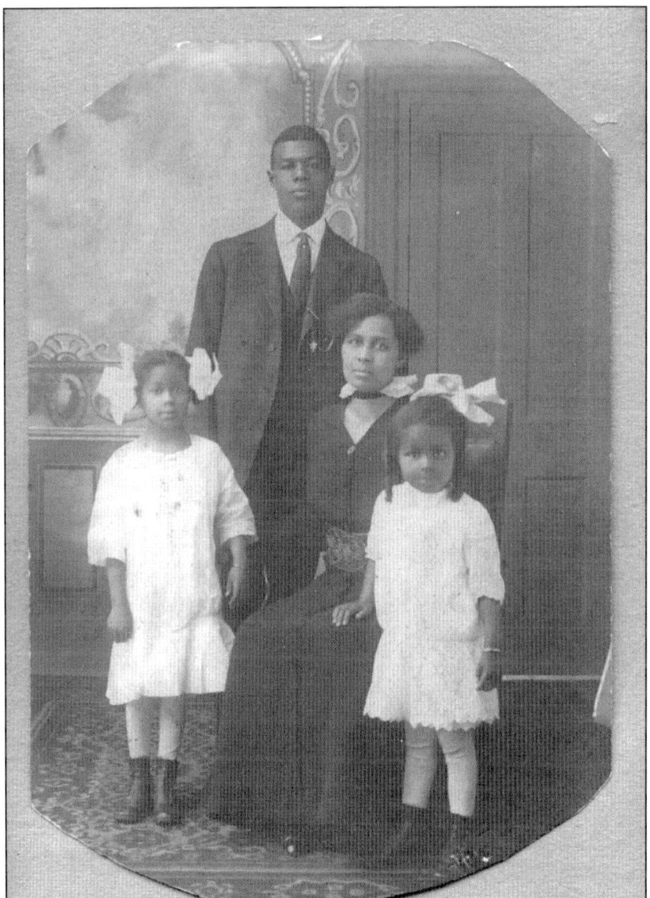

Edward Shaw and his wife, Edna, are seen here with their two daughters, Jane (left) and Lorraine, around 1920. Jane (later McPherson) was the first African American teacher in the Connellsville school system. She taught high school English from the early 1960s until her retirement in 1976. Lorraine (later Taylor) was the secretary of Payne African Methodist Episcopal Church. (Courtesy of Elsie Haley.)

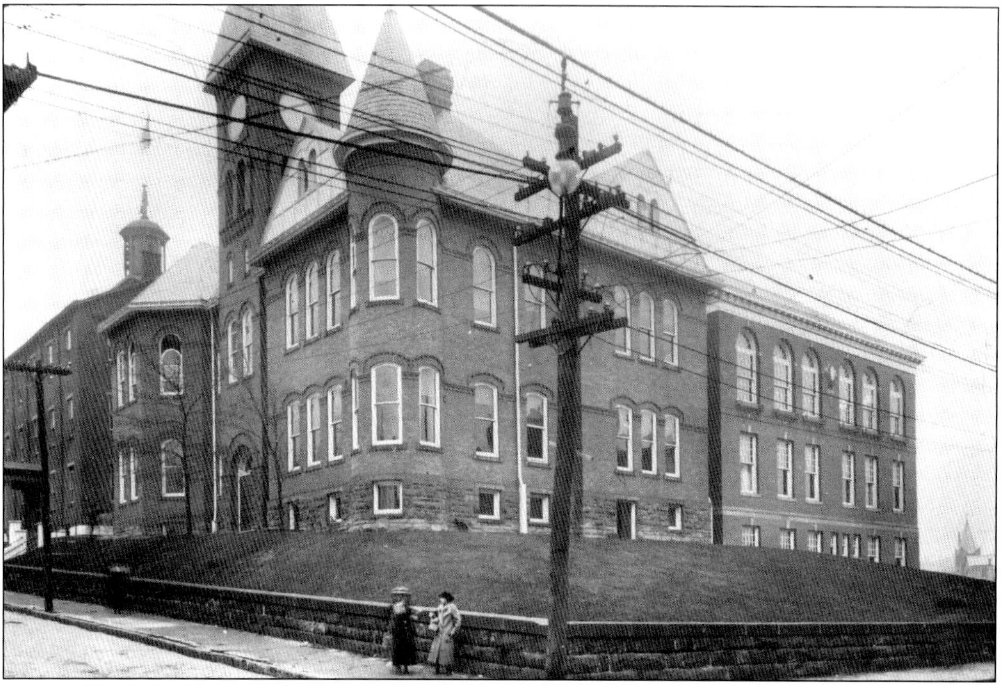

This redbrick building on the corner of East Fairview Avenue and South Pittsburgh Street was the first high school structure in Connellsville. It opened in 1895 and was replaced with a larger facility in 1916. The building was also known as the Cameron School. It continued to serve students in the junior high and elementary grades until 1970.

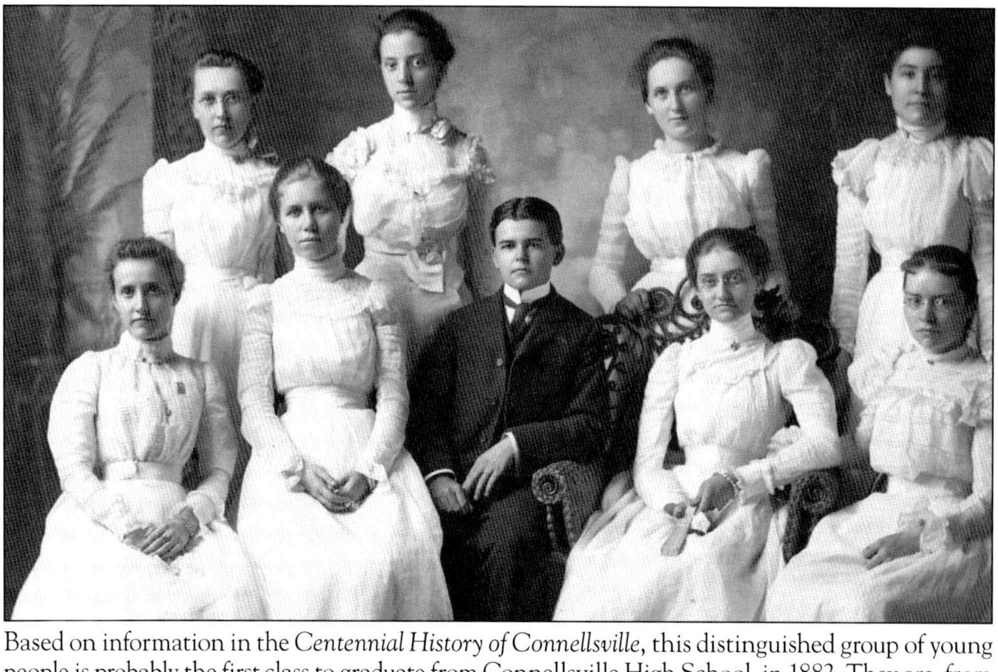

Based on information in the *Centennial History of Connellsville*, this distinguished group of young people is probably the first class to graduate from Connellsville High School, in 1882. They are, from left to right, (first row) unidentified, Sara McDivett, Theodore White, Irene Brashear, and Carrie Kenyon; (second row) Mary Kreger, Garnet Marchand, Elizabeth Belgaley, and unidentified.

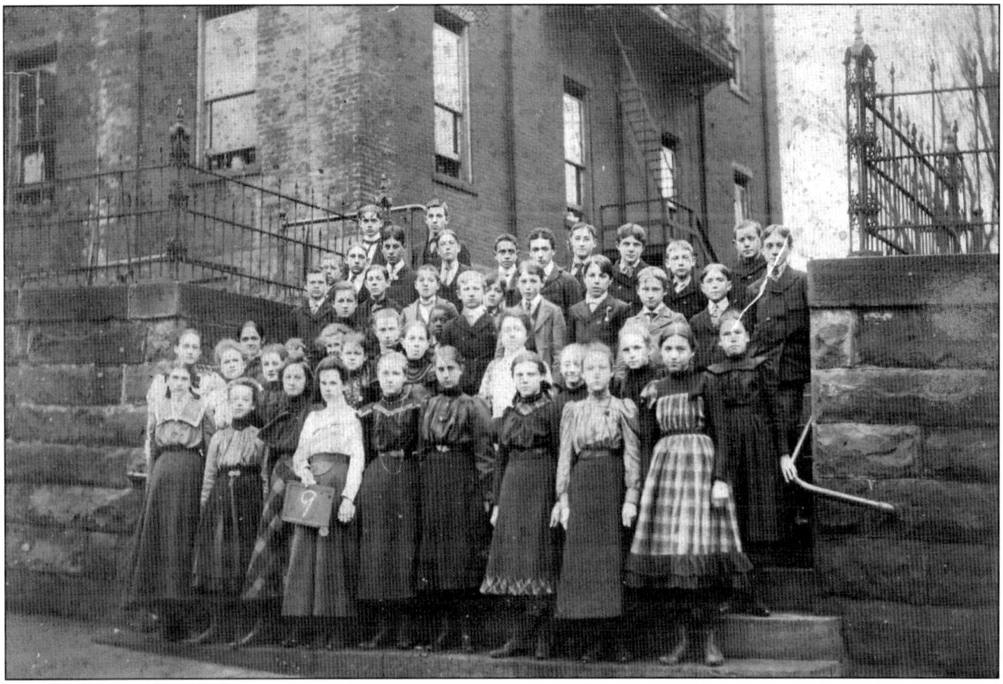

Ninth-grade students pose on the steps of the Union School on East Fairview Avenue in 1899. The first high school classes were conducted here, beginning in 1882. The school opened in 1869 and had three floors. The top level was initially used for community events such as concerts and special meetings.

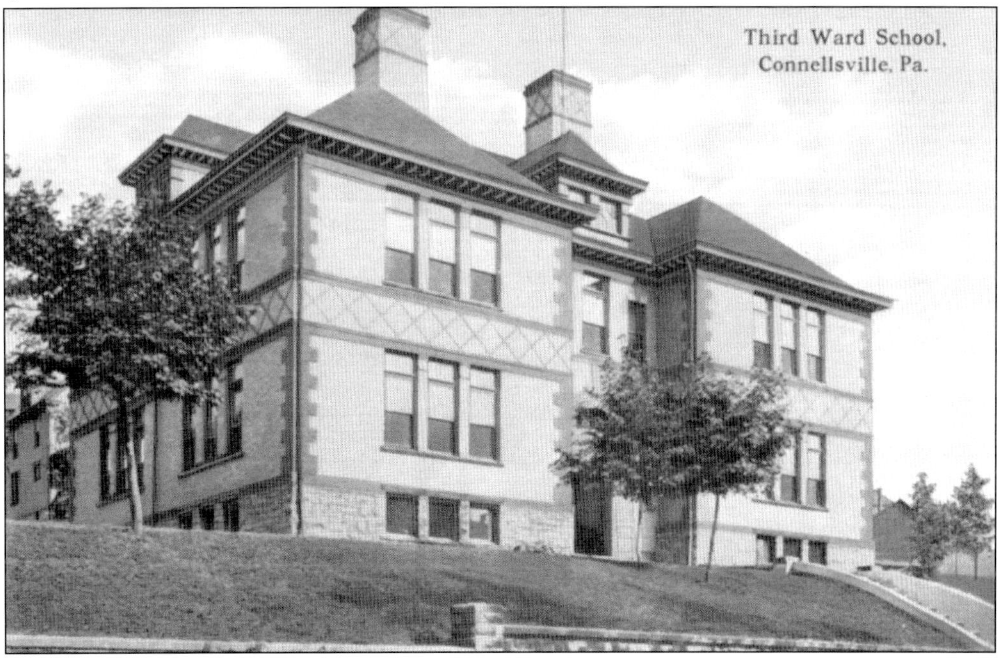

The attractive Third Ward School was built in 1900 on Jefferson Street and contained eight grades. The building, which was converted into apartments in 1974, represents a fine legacy of preserving a cherished neighborhood institution. (Courtesy of Tom Rusnack.)

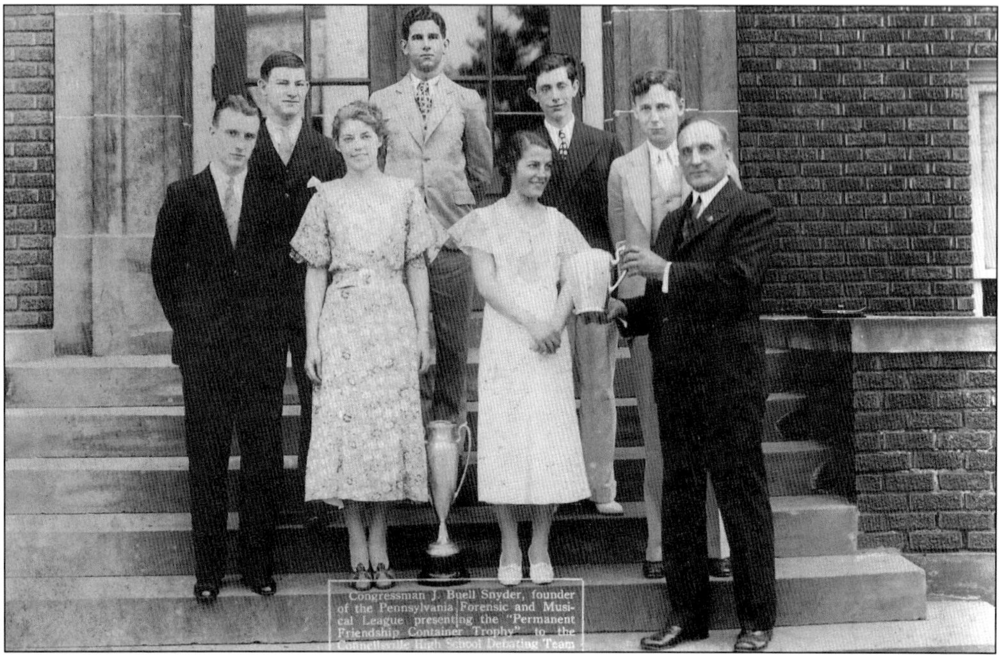

Congressman J. Buell Snyder, the founder of the Pennsylvania Forensic and Musical League, presents a trophy to Carmel Caller, representing the winning debate team from Connellsville High School. Harold Swank, a teacher and the debate team advisor, is directly to the left of Snyder. The photograph was taken in front of the high school on East Fairview Avenue in 1934.

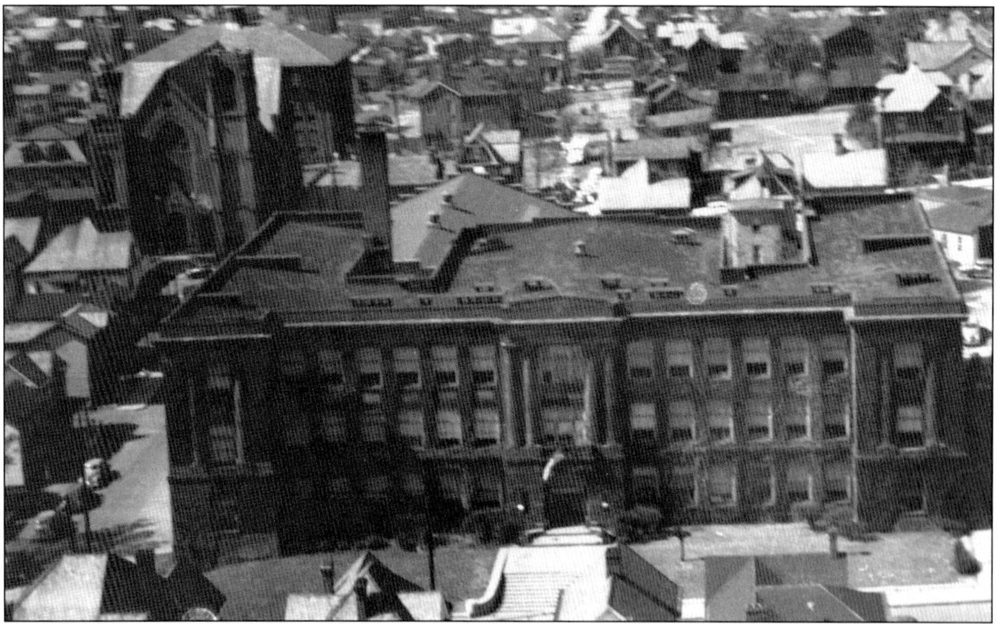

The second public high school, constructed on East Fairview Avenue, is seen here in the foreground. It operated from 1916 to 1956. The school building is now home to the Connellsville Community Center. Immaculate Conception High School is directly visible behind the parish church on North Prospect Street. Immaculate Conception was established in 1921 and closed after Geibel High School opened in 1964. The building served as a consolidated grade school until 2011.

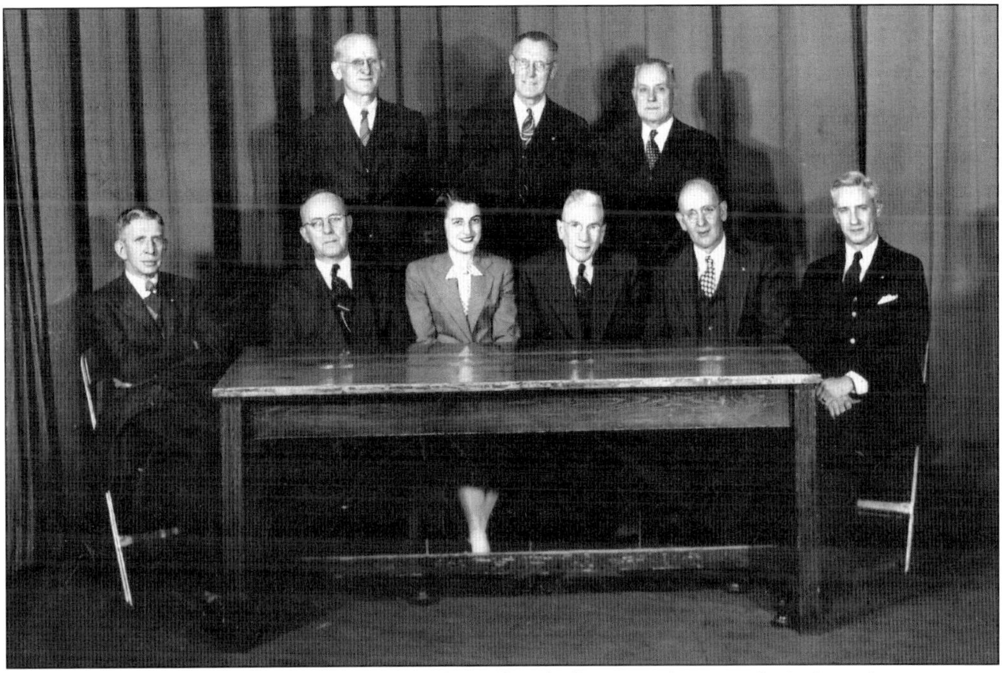

The Connellsville School Board sat for this official photograph around 1940 in their room at the high school on Fairview Avenue. They are, from left to right, (seated) Dr. Clyde Campbell, Charles Payne, secretary Carmel Caller, Fred Munson, J.R. Mestrezat, and solicitor Donald Highbee; (standing) D.H. McIntyre, W.L. Zollars, and superintendent W.G. Davis.

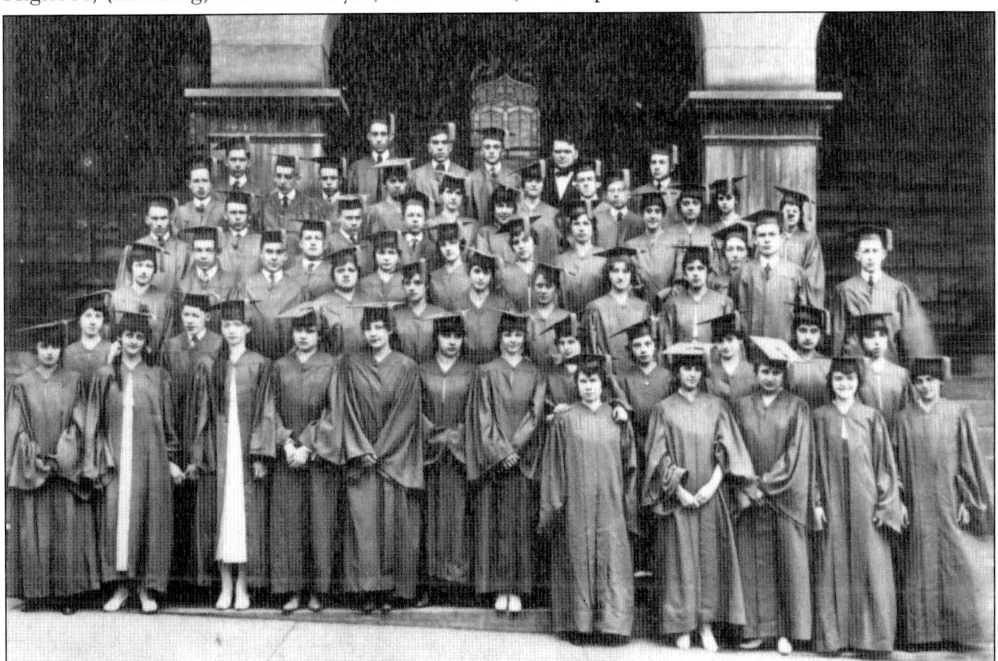

The 1915 graduating class, the last to graduate from the first high school building, known as the Cameron School, posed for this photograph in front of Trinity Lutheran Church on East Fairview Avenue.

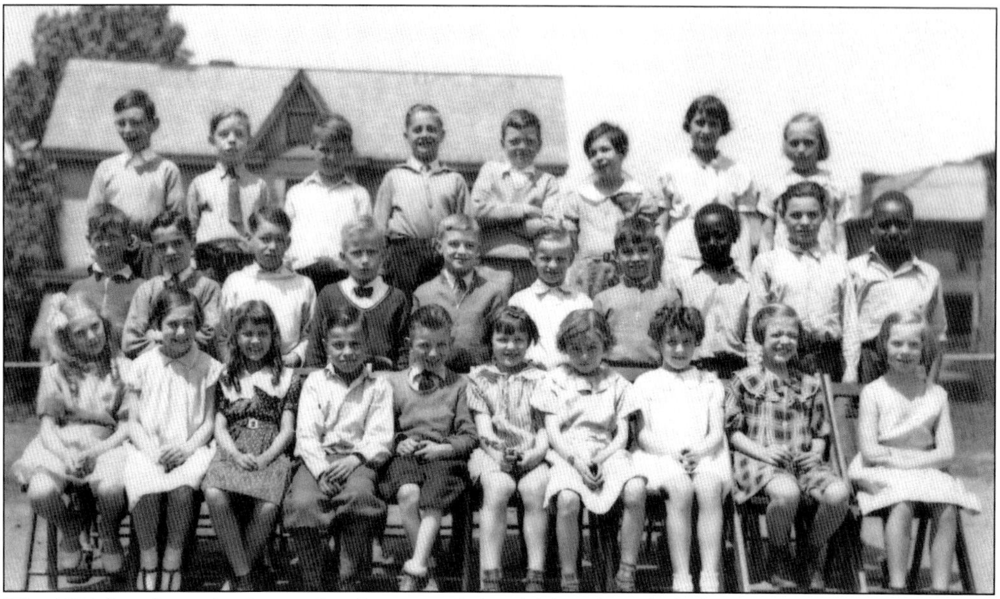

The 1934 fourth-grade class at the Cameron School (above) included, from left to right, (first row) Mildred McNair, Betty Lou Lewis, Vivian Calhoun, Donald Small, Clyde Smith, Rose Moser, Donna Camlin, Greta Piper, Marie Conniff, and Delores Schrock; (second row) Robert Miner, John Nardine, Harry Prytulak, Bert Swartzwelder, Jack Martin, Wilford McClintock, Jerry Evans, unidentified, Clarence Miller, and Clifford Ewell; (third row) Billy McFarland, Wilbert Woodward, Robert Miner, Harold Frazier, Robert Darrell, Virginia Morris, and two unidentified students. The Crawford School, seen below on a 1940s postcard, was on Seventh Street and served elementary students on the Westside. After the Zachariah Connell Grade School opened on Park Street, the building housed the district's administrative offices until 2006. (Above, courtesy of Bert Swartzwelder; below, courtesy of Barbara and Tony Keefer.)

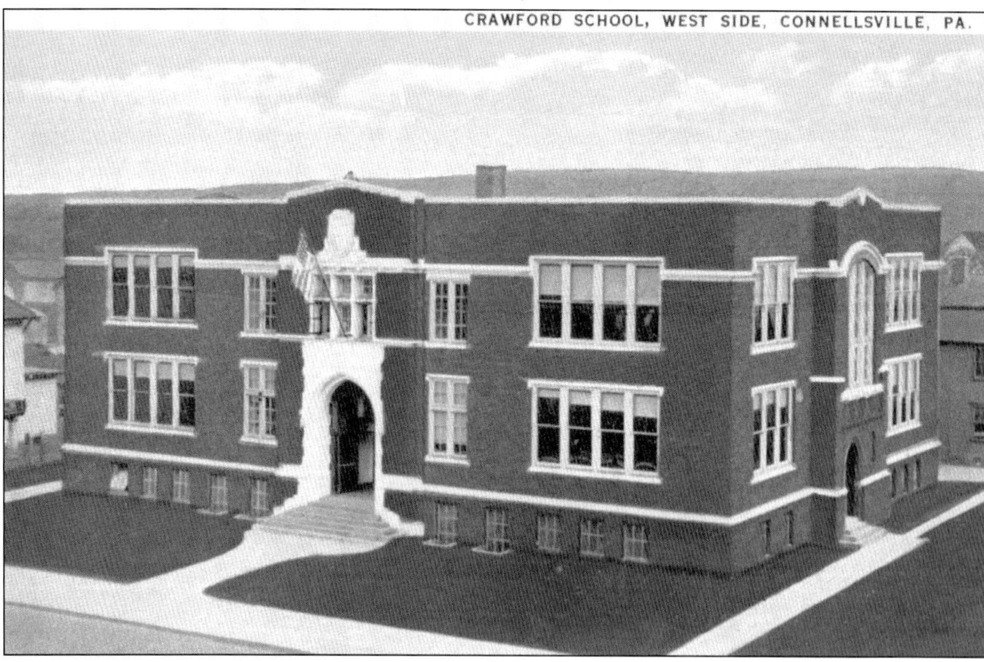

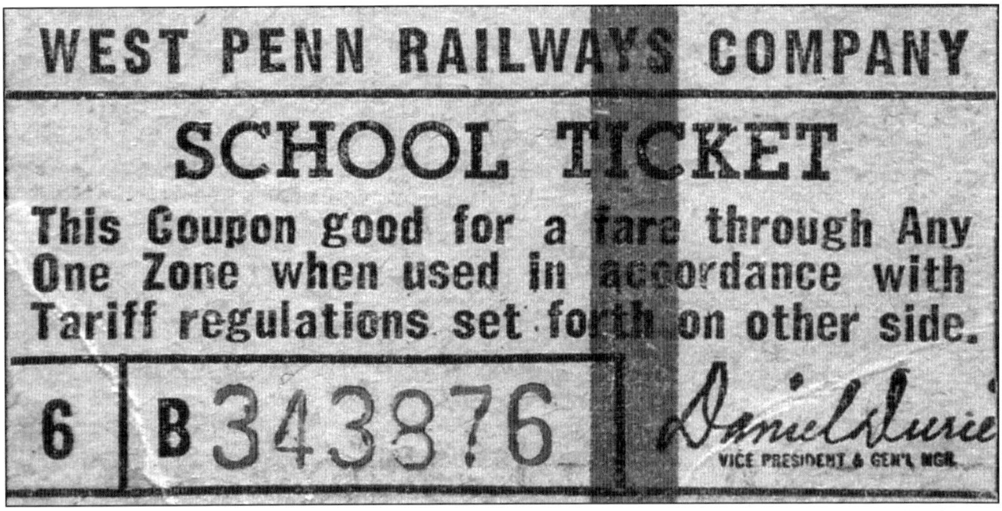

Before school districts merged in the 1950s, students usually walked to neighborhood schools. However, some lived far enough away that they needed transportation, and the West Penn offered reduced fares for students, as seen on this ticket. (Courtesy of John Husband.)

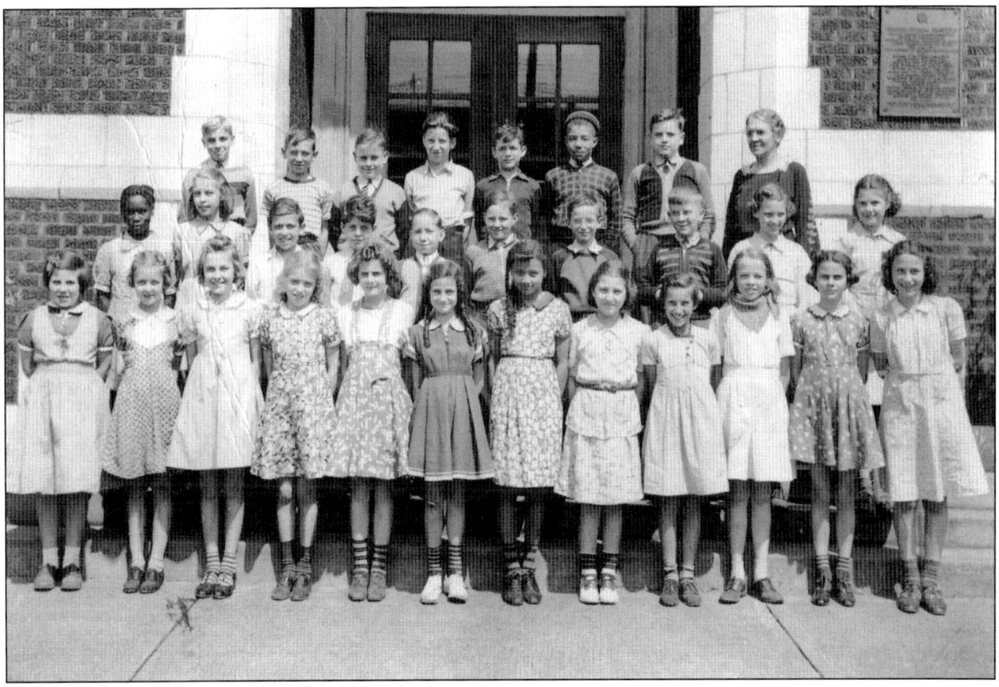

The 1938 sixth-grade class at the Crawford School is seen here. Rosemary Capo Tressler is in the first row on the far right. The teacher is a Mrs. Woods. The other students are unidentified. (Courtesy of Sandy Russell.)

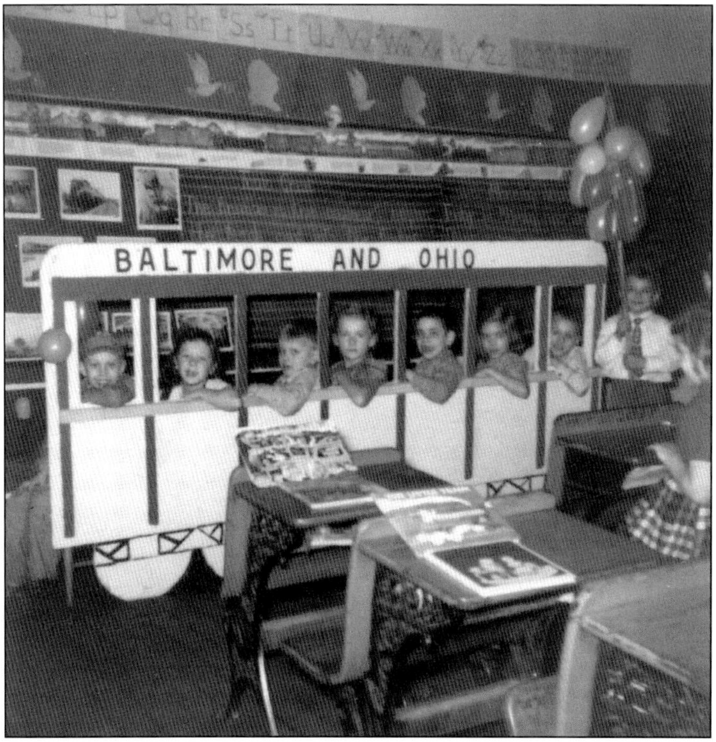

Elementary students at the Gibson School in South Connellsville are seen here in 1944 as they complete a study unit on transportation that focused on the B&O Railroad. The exercise included a play in which class members "rode" on the railroad. The Gibson School was on South Pittsburgh Street. When the school closed, students were transferred to the nearby Southside Grade School. (Courtesy of Pete Maracondi.)

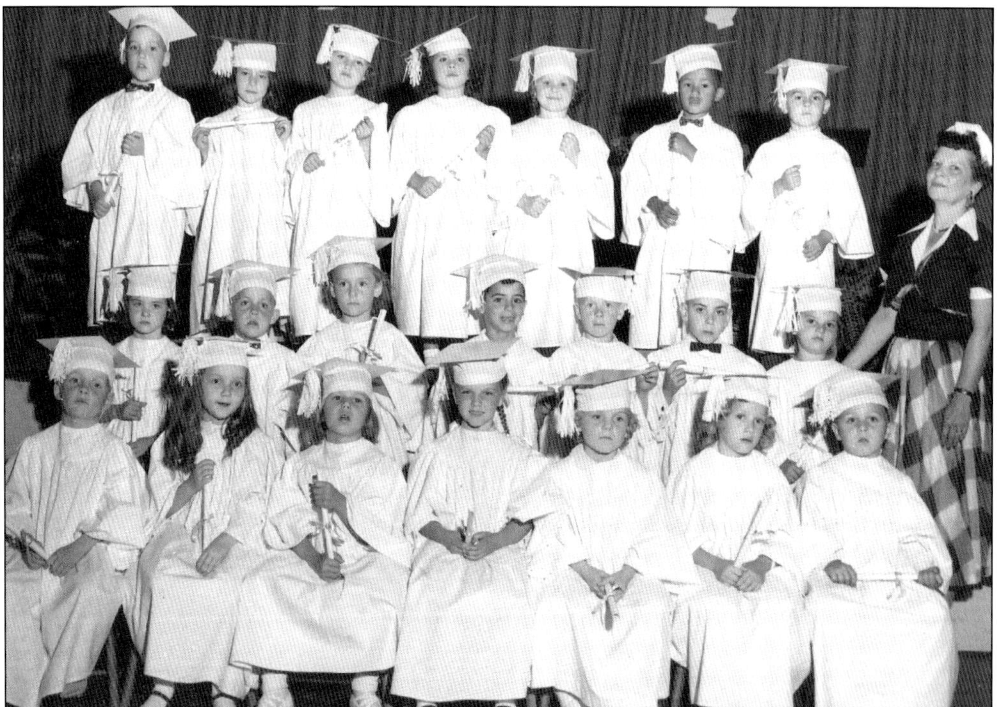

Gertrude Whipkey, a highly respected teacher at the Gibson School, poses with her students at their kindergarten graduation in the mid-1950s. (Courtesy of Pete Maracondi.)

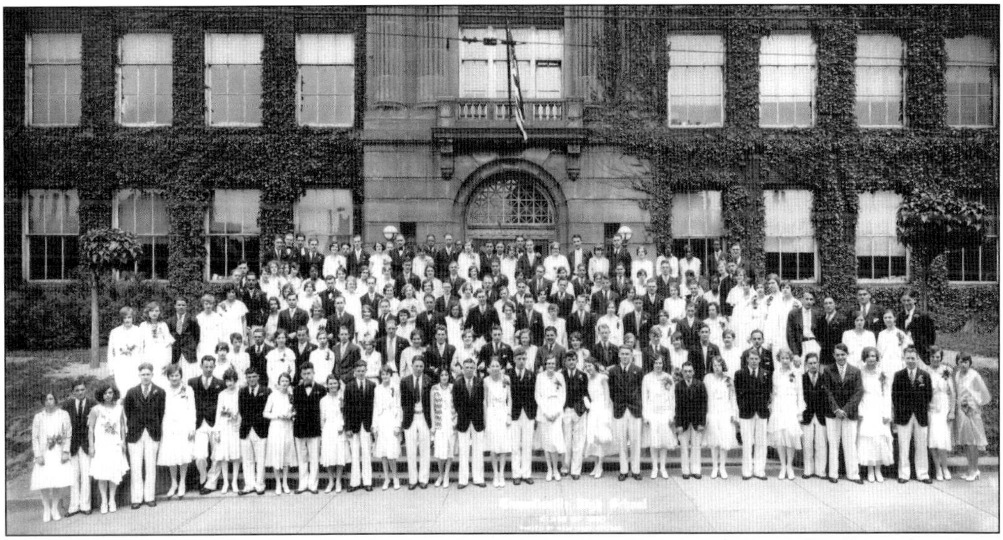

The 1930 graduating class assembled for this photograph on the steps of Connellsville High School on Fairview Avenue on Senior Day, a special day that featured ceremonies and a luncheon at one of Connellsville's many restaurants or private clubs. It was also customary for young women to wear white dresses for the occasion, while young men dressed in suits or other formal attire. (Courtesy of John Husband.)

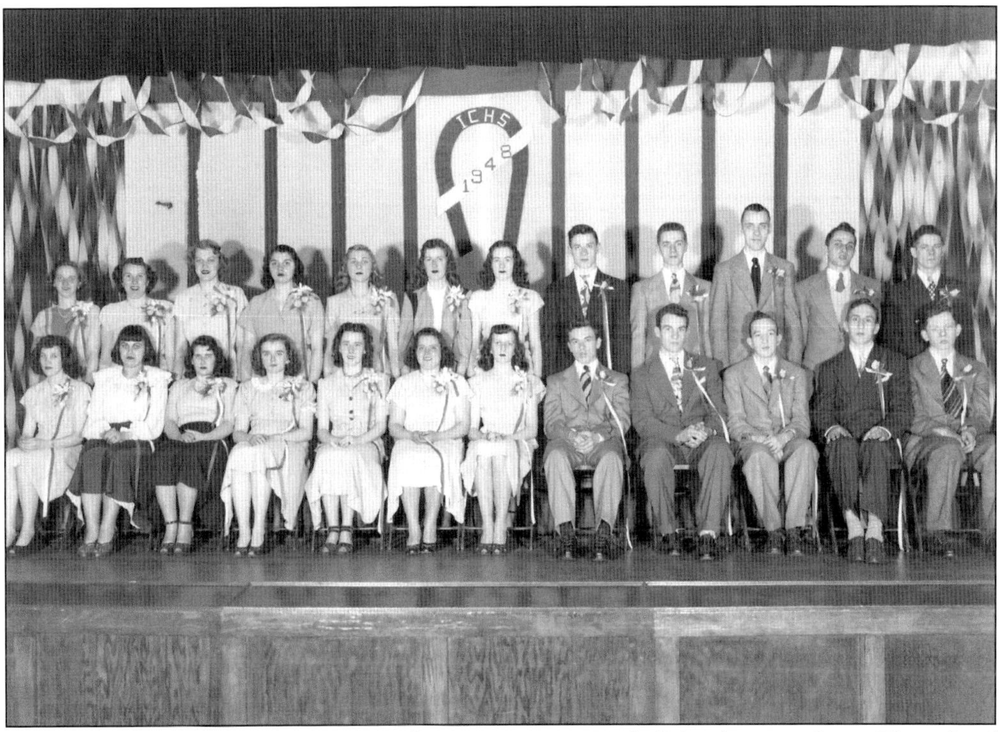

The 1948 graduating class at Immaculate Conception High School is seen here. The school started with a two-year curriculum in 1921 and was expanded to four years under the pastorate of Rev. Henry Geibel. The first students to complete the four-year program graduated in 1930. (Courtesy of John Ed Dailey.)

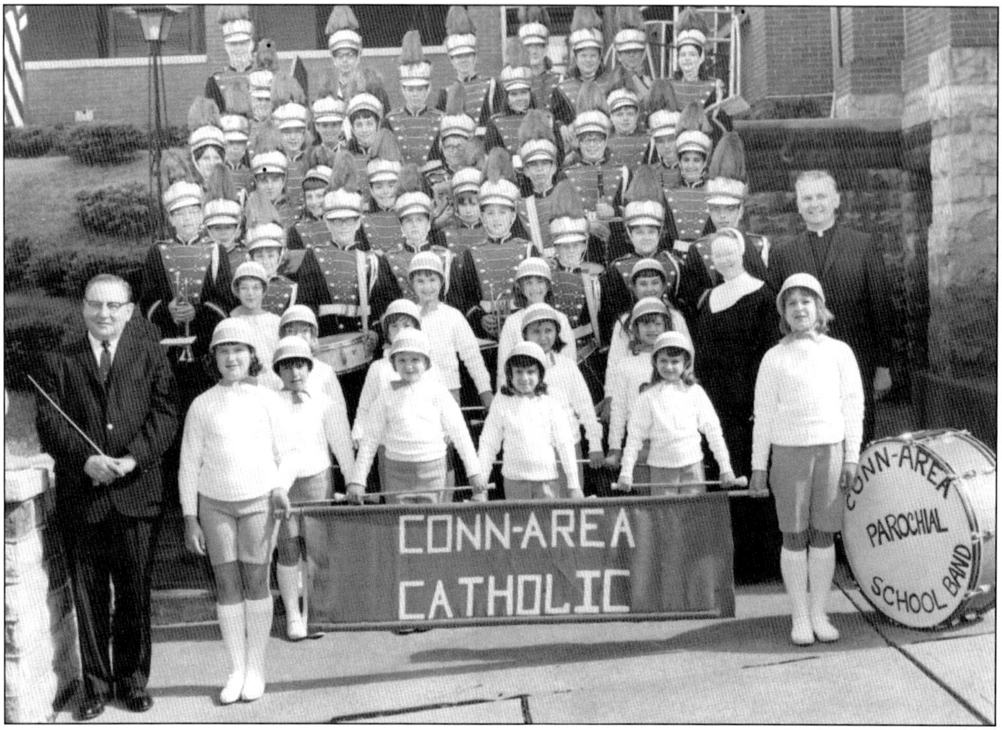

The Connellsville Area Catholic Grade School band assembled on the steps of the Immaculate Conception rectory for this photograph in the late 1960s. Standing in the first row on the far left is band director Carmine Molinaro Sr. The priest is Reverend Nyeste. The band uniforms were donated by the Connellsville School District after the school colors were changed from orange and black to blue and white. (Courtesy of Margaret Molinaro.)

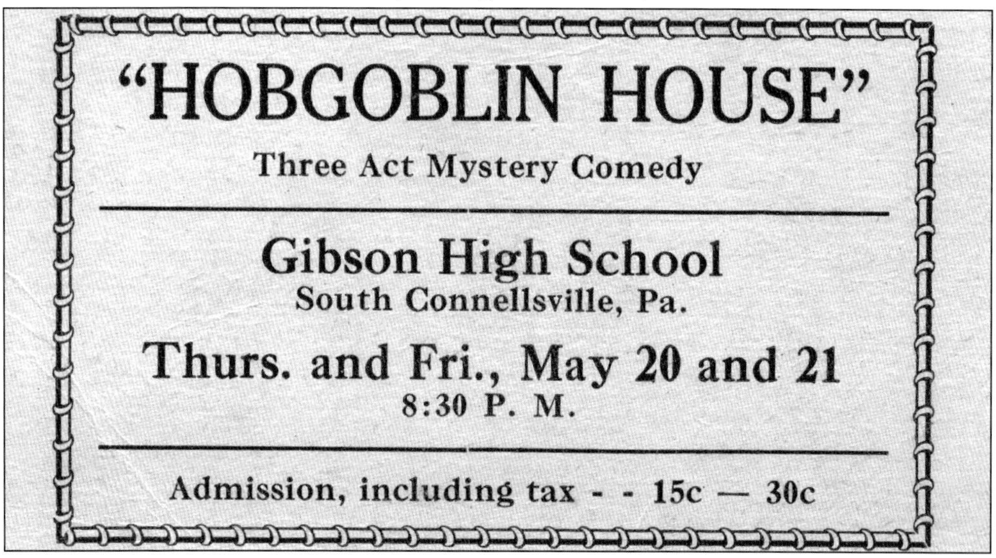

The sophomore class at Gibson High School performed this three act play in May 1943. South Connellsville supported a two-year high school until 1946. After 1946, freshmen and sophomores transferred to Connellsville High School for their secondary education. (Courtesy of Pete Maricondi.)

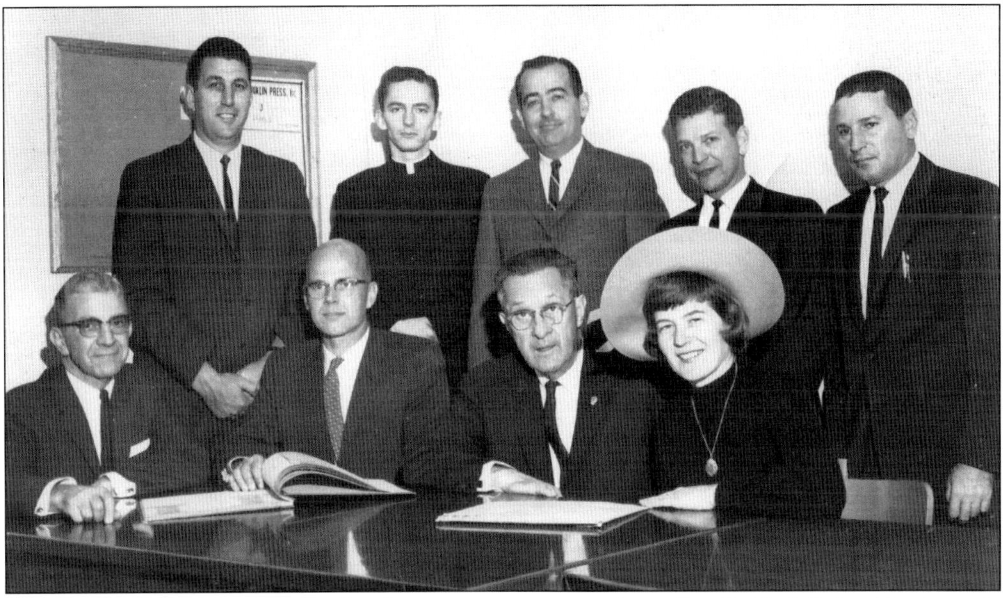

Pennsylvania State University offered extension courses at Connellsville High School in the early 1960s before establishing a campus on Route 119 in Dunbar Township. The first trustees were, from left to right, (seated) unidentified, John Shallenberger, and two unidentified; (standing) John "Wally" Schroyer, Reverend Sheridan, unidentified, Clifford Pritts, and Richard Carpinelli. (Courtesy of John "Wally" Schroyer.)

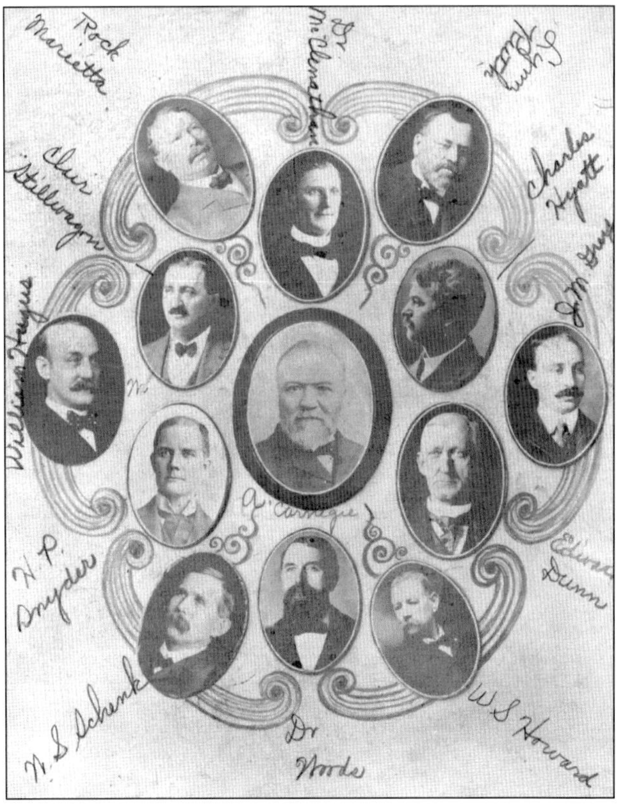

The members of the first library board of trustees are seen here in 1904, surrounding Andrew Carnegie in the center. Board president Dr. J.C. McClenathan is directly above Carnegie, followed clockwise by Linford Ruth, Charles Hyatt, J.M. Grey, Edmund Dunn, Samuel Howard, Dr. Samuel Woods, Winfred Schenck, H.P. Synder, William Hugus, Clair Stillwagon, and Rockwell Marietta. Carnegie appointed six of the board members himself. Other trustees were selected from the school board and the borough council.

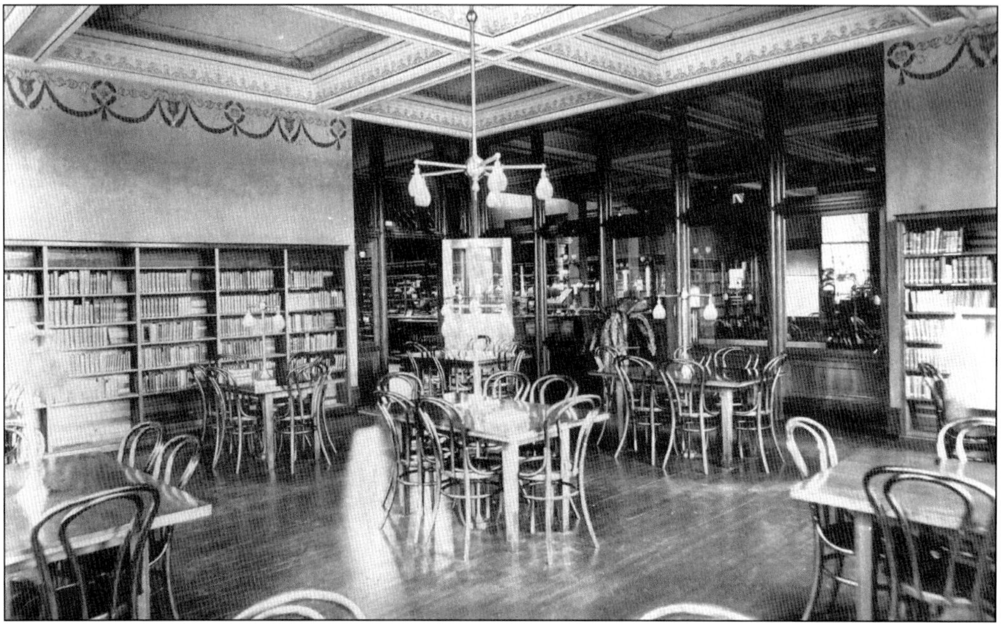

One of the reading rooms at the Carnegie Library is seen here shortly after it opened to the public on May 1, 1903. Andrew Carnegie donated $50,000 for its construction, which was subsequently followed by a donation of $18,000 to improve the grounds surrounding the facility. In return for these donations, the school board and the borough council agreed to appropriate funds for operational expenses. The library was erected on the site of the old Connell graveyard. (Courtesy of Felix Prestia.)

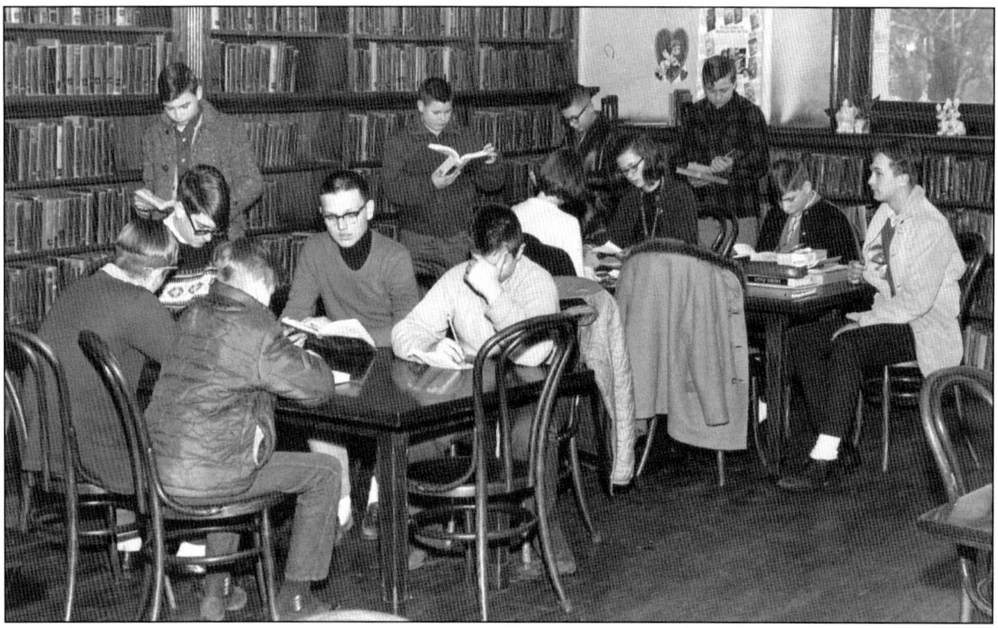

When the new high school on Fairview Avenue was designed, the plans did not include a library. It was reasoned that students could use the nearby Carnegie Library as an alternative. Here, a junior high school class uses the library's Getty Room in 1960. Students could make the trip to and from the high school within one class period. (Courtesy of Tom Korba.)

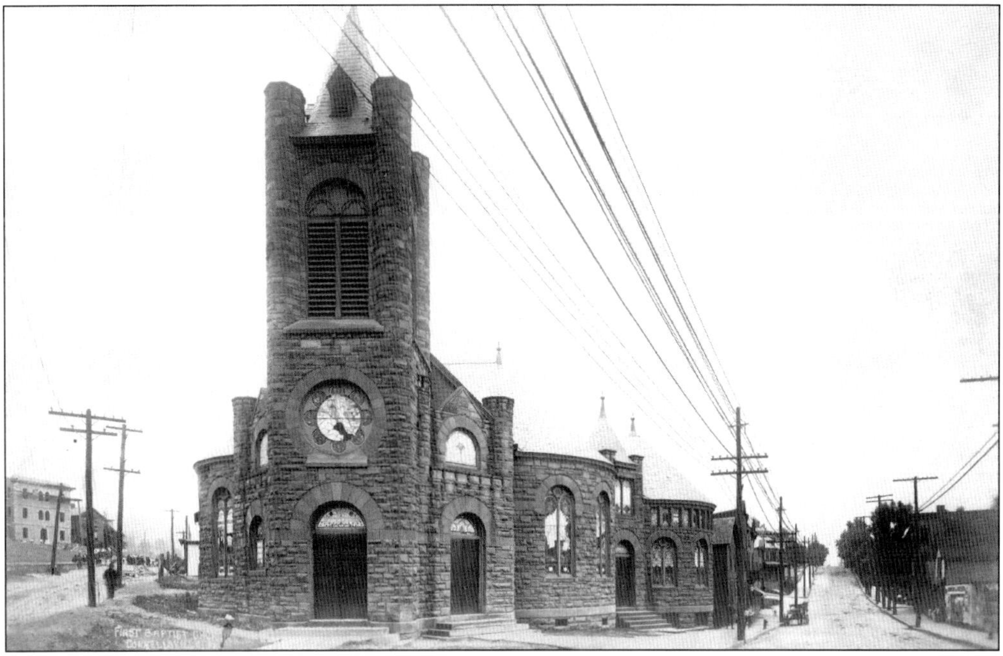

Sam and Caleb Trevor came to Connellsville in 1795 and donated land for the First Baptist Church on the corner of Pittsburgh and Apple Streets in 1817. The original building was replaced by a larger structure on the same property in 1877. As the congregation continued to expand, members erected this house of worship at the intersection of Baldwin Avenue and South Pittsburgh Street in 1903.

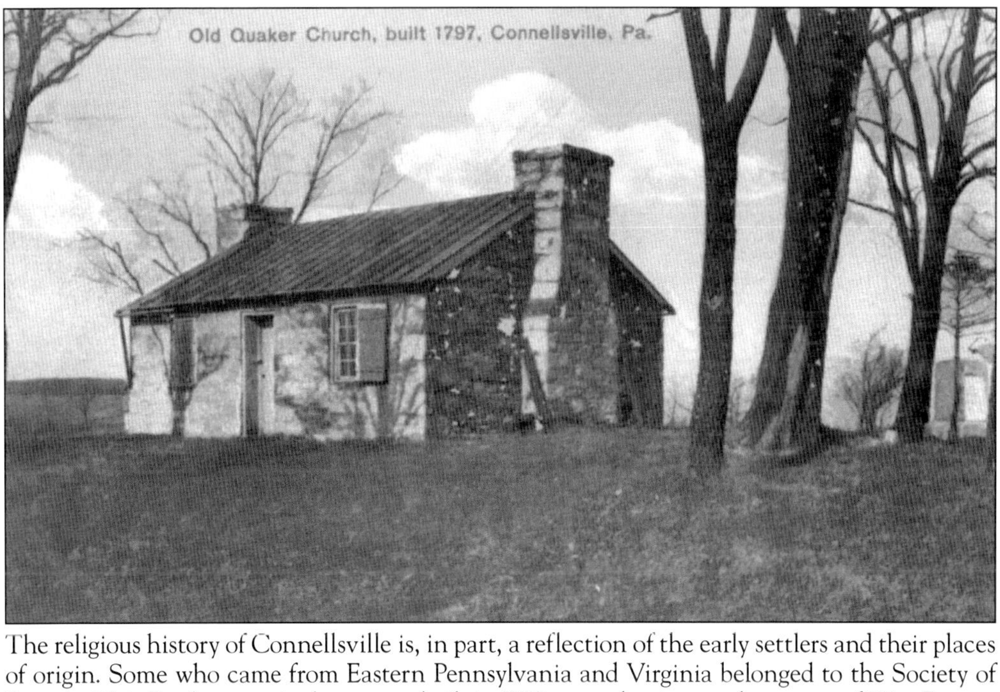

The religious history of Connellsville is, in part, a reflection of the early settlers and their places of origin. Some who came from Eastern Pennsylvania and Virginia belonged to the Society of Friends. This Quaker meetinghouse was built in 1797 near what is now the corner of West Fayette Street and York Avenue. (Courtesy of Tom Korba.)

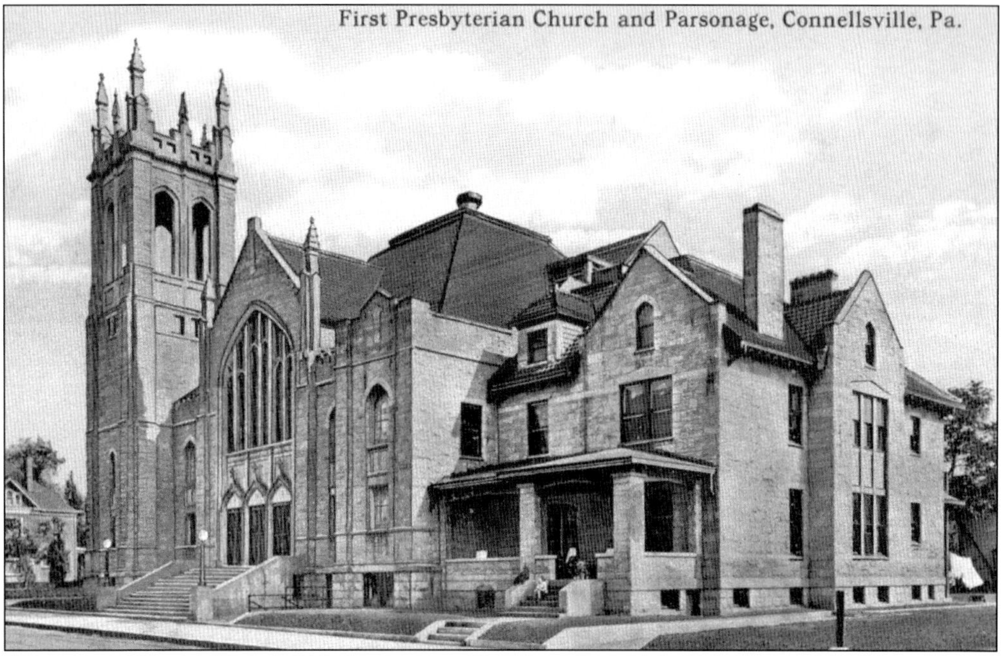

Connellsville Presbyterian Church was organized in 1831. Initially, members worshipped in a church on Crawford Avenue. The congregation then moved to the imposing stone edifice above on South Pittsburgh Street in 1916. The church is recognized throughout the region for its beautiful stained-glass windows, which were manufactured and imported from Germany.

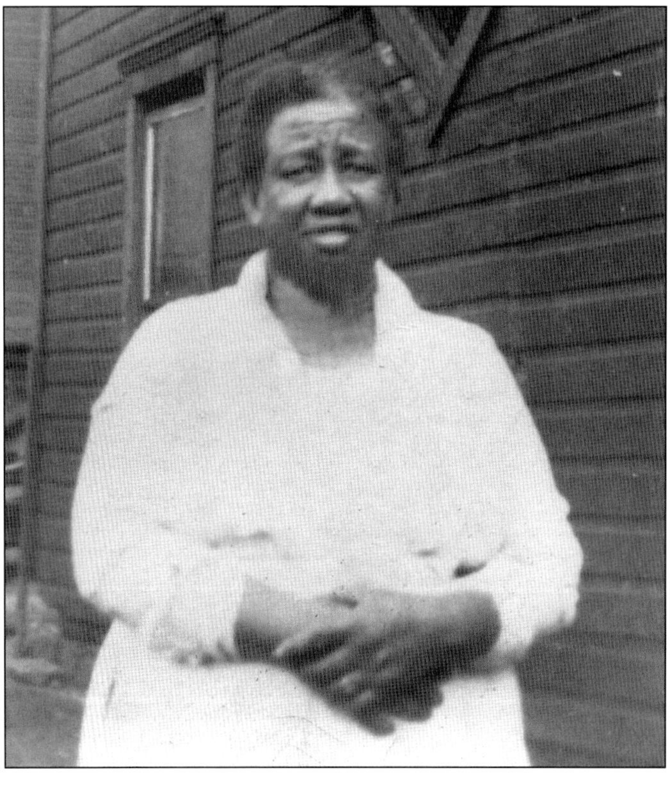

Annie Fox Ennis is seen here in the early 1900s. She was an active member of Payne African Methodist Episcopal Church, which was organized in 1875. Ennis was the first secretary of the congregation. The current church, located on West Crawford Avenue, was completed in 1881. (Courtesy of Elsie Haley.)

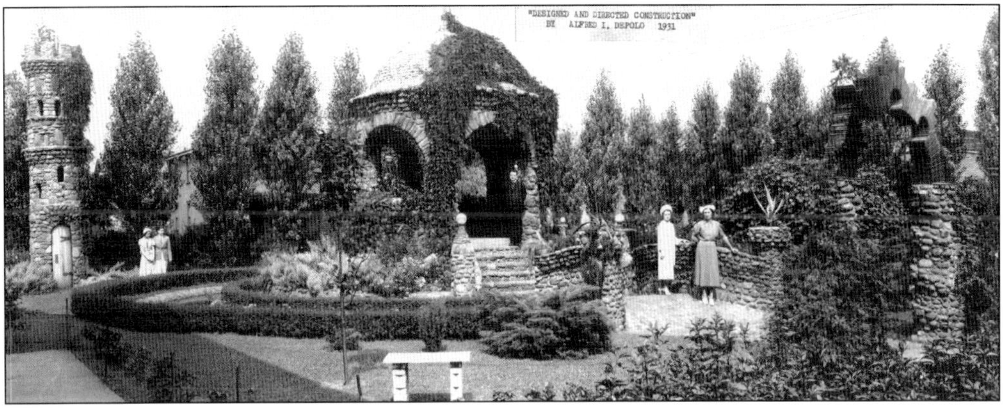

This grotto was constructed during the Great Depression by unemployed parishioners from St. Rita's Parish, which was founded by Italian Catholics in 1915. The garden, situated behind the church on South First Street, was dedicated in 1933. Alfred De Palo designed and supervised the construction. The stones were obtained from the Youghiogheny River.

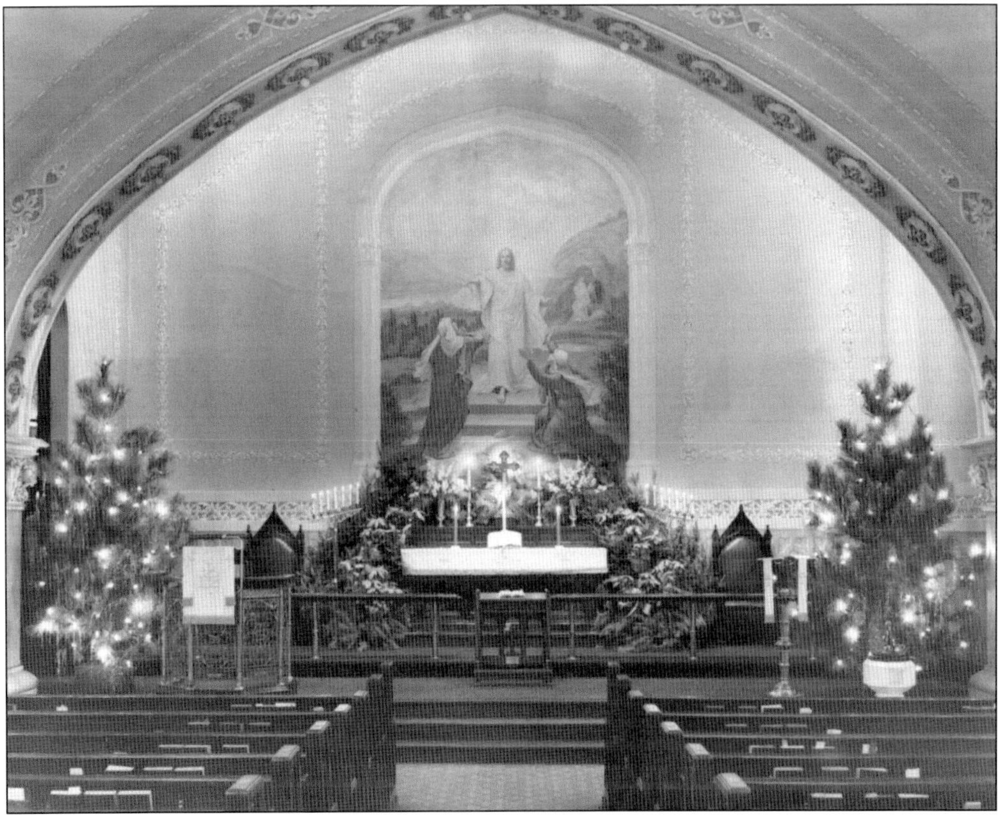

Trinity Lutheran Church, on the corner of East Fairview and Carnegie Avenues, features this mural gracing the sanctuary. The area is illuminated by an arch containing electric lights. This photograph was taken during the Christmas season in the 1930s. While this church was erected in 1910, the congregation was organized in 1881 and built their first house of worship on Apple Street.

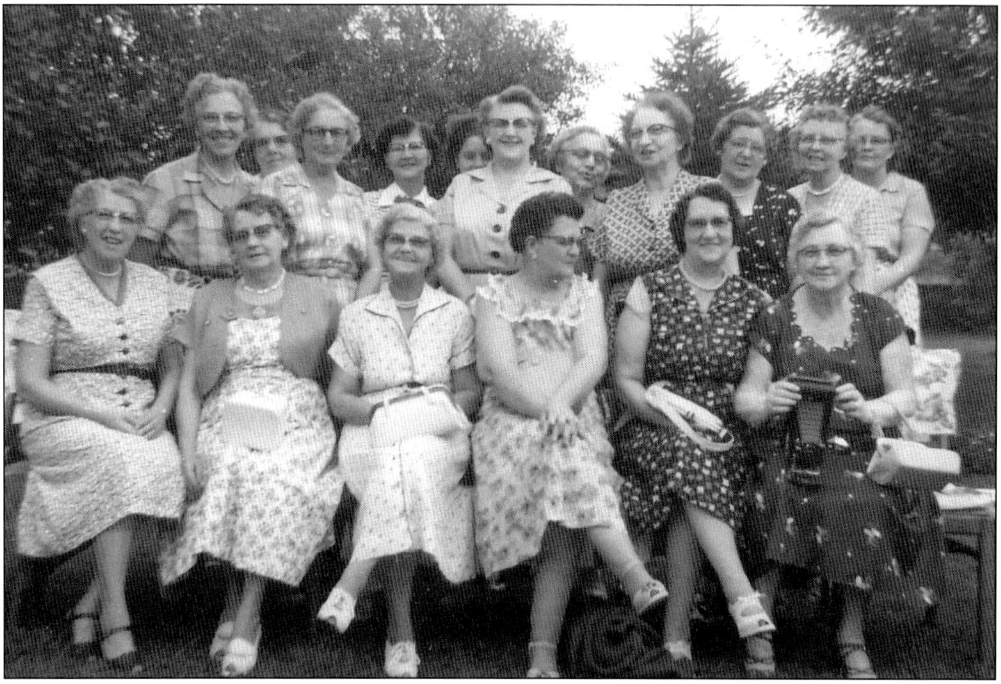

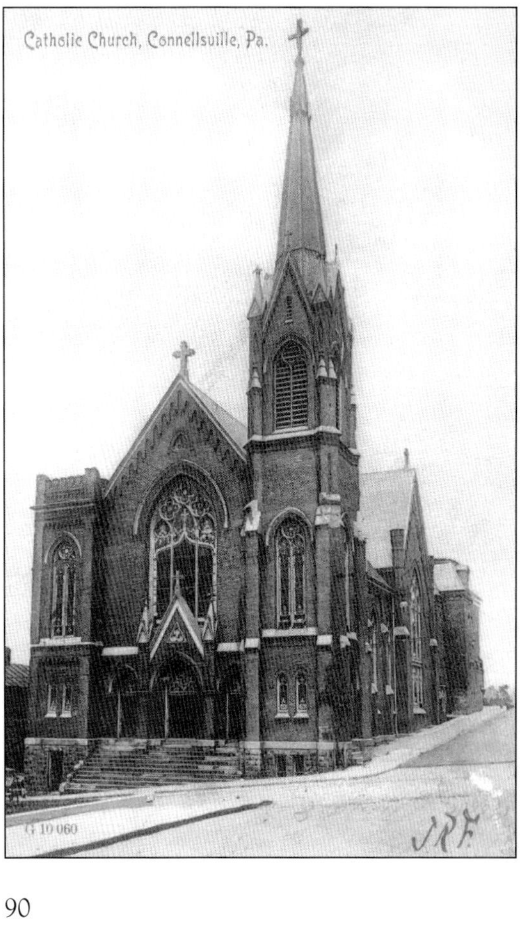

Members of the Lottie A. Bradley Sunday school class at the Albright Evangelical United Brethren Church in South Connellsville posed for the photograph above in 1950. They are, from left to right, (first row) Elizabeth Tressler, Esta Coughenour, Olive Lee, Vella Grieves, Edna Nelson, and Dessie Travis; (second row) ? Bates, ? Bloom, Dolly Burke, two unidentified, Ida Cox, Rhea Winkler, Tillie Sanner, Gertrude Lincoln, ? Stewart, and Ruth Martz.

Immaculate Conception Church, built in 1897, was the anchor for parish facilities that occupied an entire block, from East Crawford Avenue to Apple Street. The church replaced an earlier structure destroyed by fire in 1892. As seen here, the new church originally had a steeple, but it was removed in the 1930s after high winds created structural issues. (Courtesy of Felix Prestia.)

Connellsville Christian Church, also known as the Church of Christ or the Disciples of Christ, was organized in 1832. The congregation occupied two other churches before they built this current edifice in 1892 on South Pittsburgh Street. According to the *Centennial History of Connellsville*, one of the denomination's founders, Thomas Campbell, and his son Alexander preached to the local congregation in the mid-1800s.

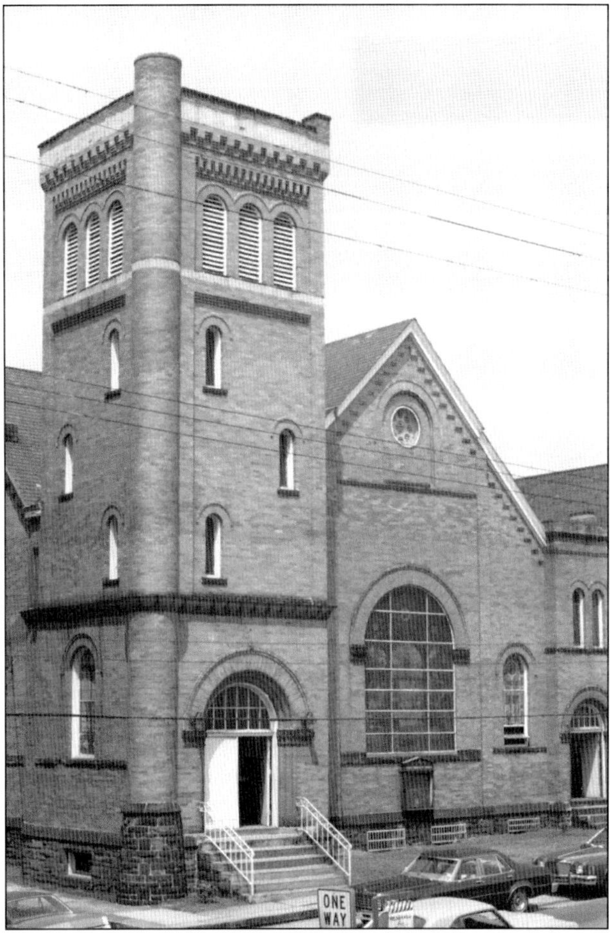

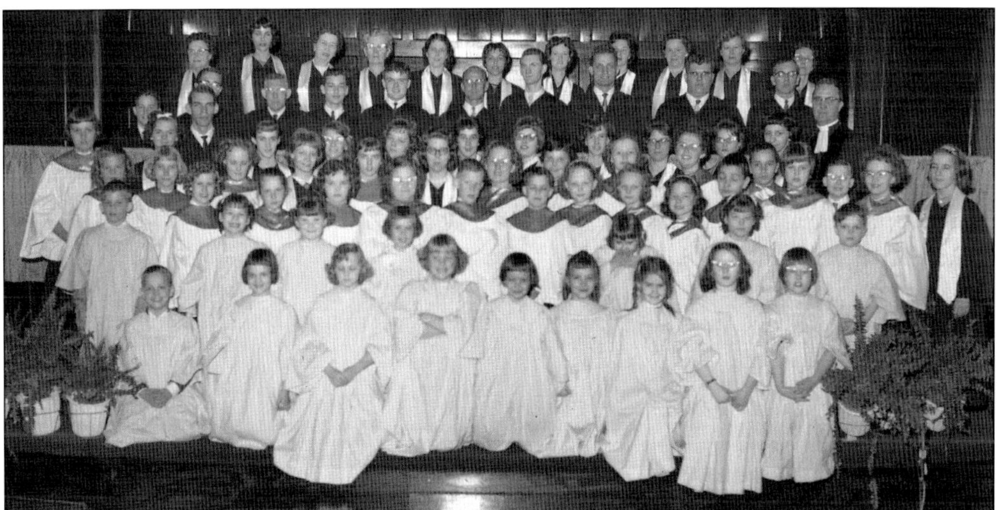

Three separate groups, consisting of young people, teenagers, and adults, combined to form this remarkable choir at Connellsville United Presbyterian Church in 1961. The choir director was Paul Curlee and the minister was Dr. Francis Stewart. (Courtesy of Bill Sechler.)

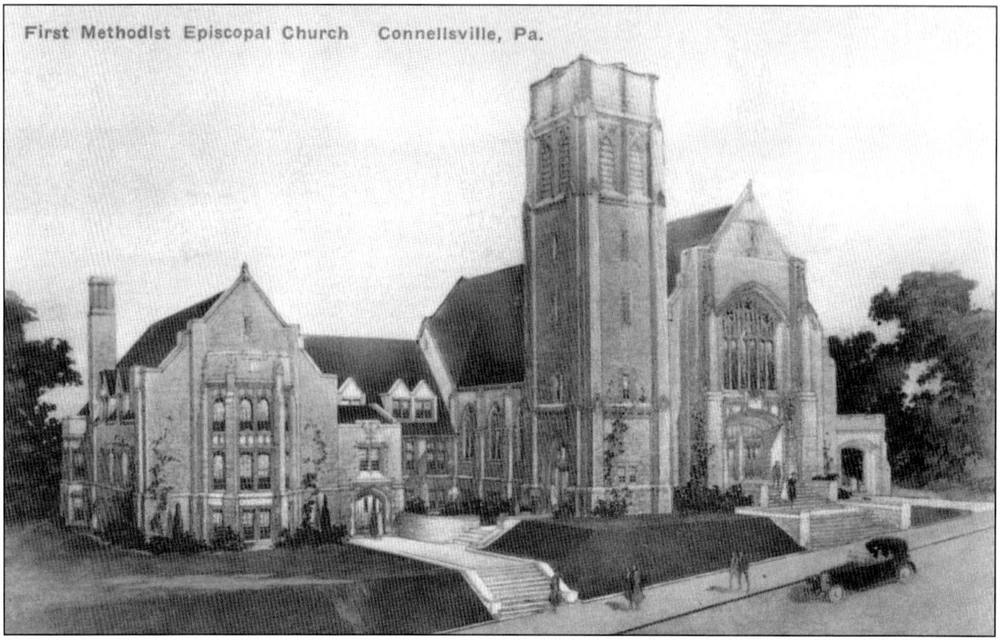

The Methodist Society in Connellsville was formed around 1830. They initially worshipped on the corner of North Arch and West Apple Streets until they constructed the imposing church above on South Pittsburgh Street in 1927. The property was originally owned by Linford Ruth, a prominent civic leader whose home is seen below. Ruth was the president of Colonial Bank and the Title & Trust Company of Western Pennsylvania. He helped organize the YMCA and was an original trustee of the Carnegie Library. (Above, courtesy of Sandy Russell; below, courtesy of Felix Prestia.)

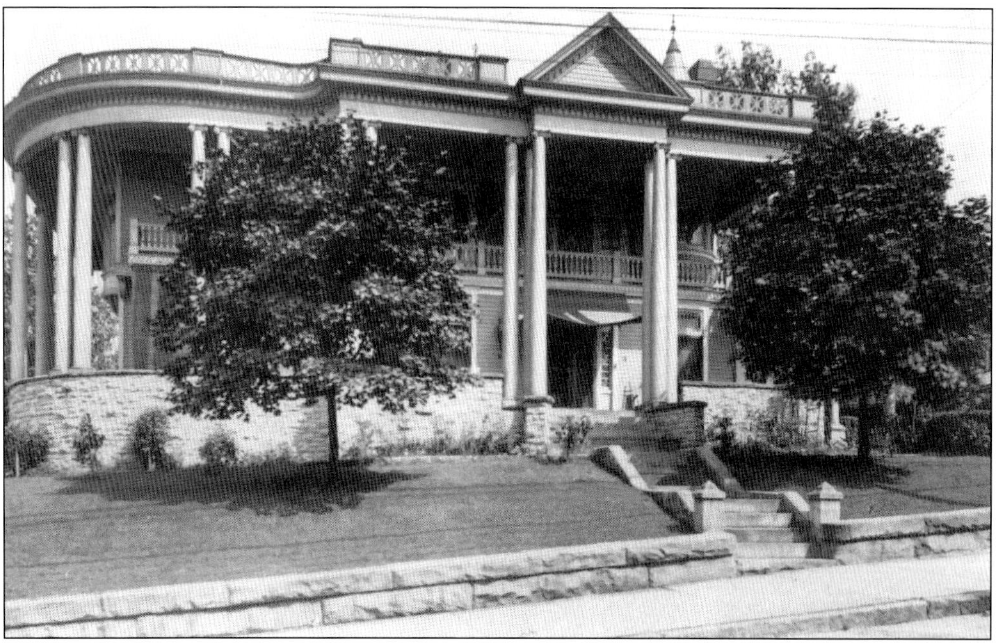

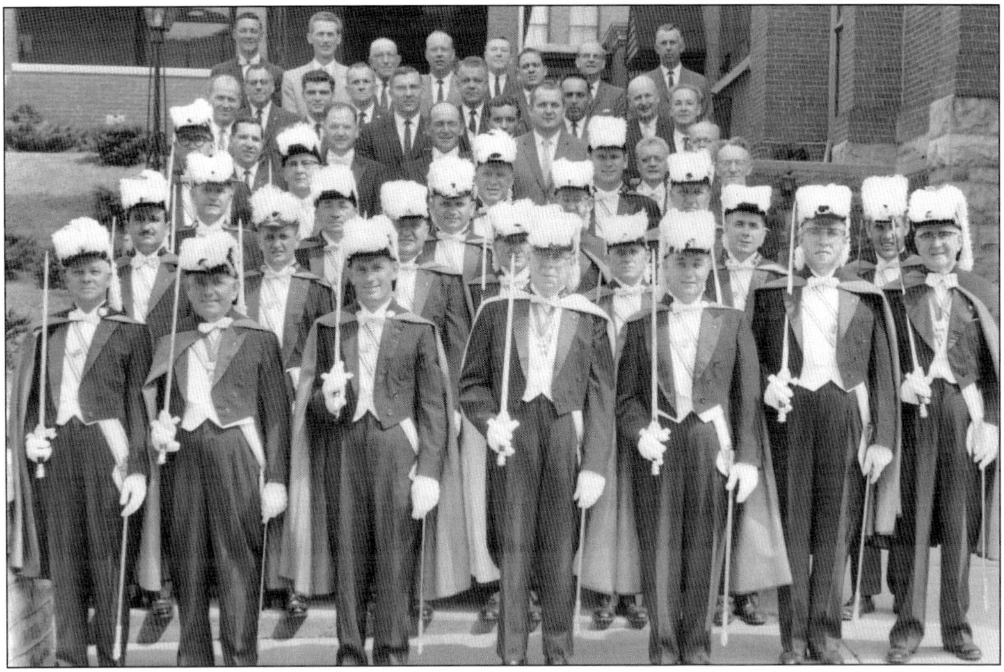

Knights of Columbus Council No. 984 assembled for this photograph on the steps leading to the rectory of Immaculate Conception Parish after a communion breakfast on May 31, 1965. The individuals in formal attire are members of the Bishop O'Connell Assembly of the Fourth Degree. (Courtesy of John Ed Dailey.)

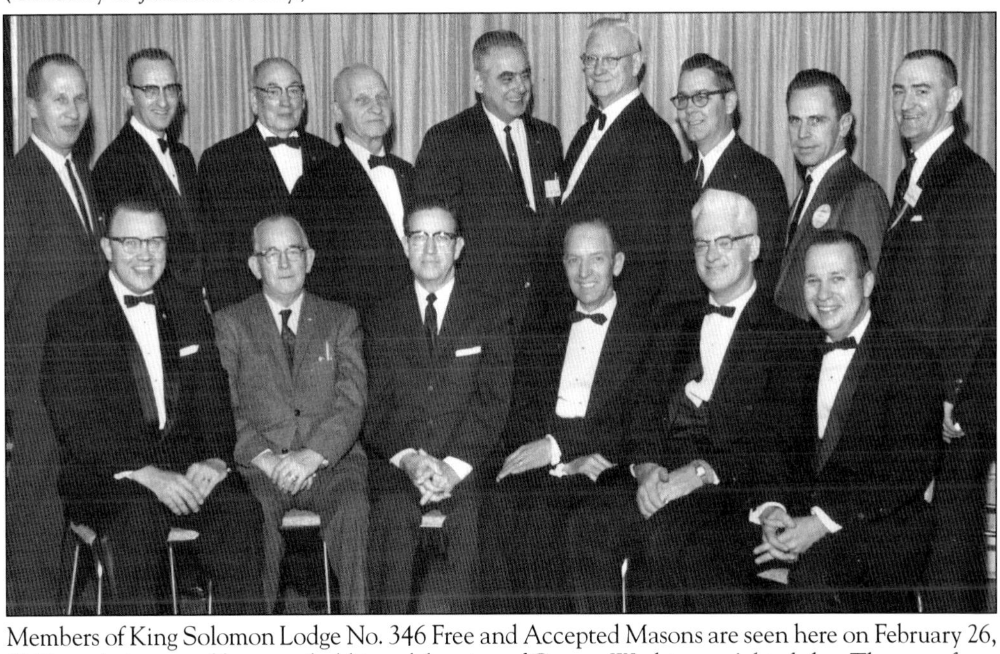

Members of King Solomon Lodge No. 346 Free and Accepted Masons are seen here on February 26, 1966, at their annual banquet held in celebration of George Washington's birthday. They are, from left to right, (first row) Louis Wandel, Harry Mitchell, three unidentified, and Vernon Stanbaugh; (second row) Robert Sandusky, two unidentified, Paul Sandusky, C.J. Franklin Ellis, unidentified, Perry Culver, Jonas Hammaker, and Charles Priest. (Courtesy of King Solomon Lodge.)

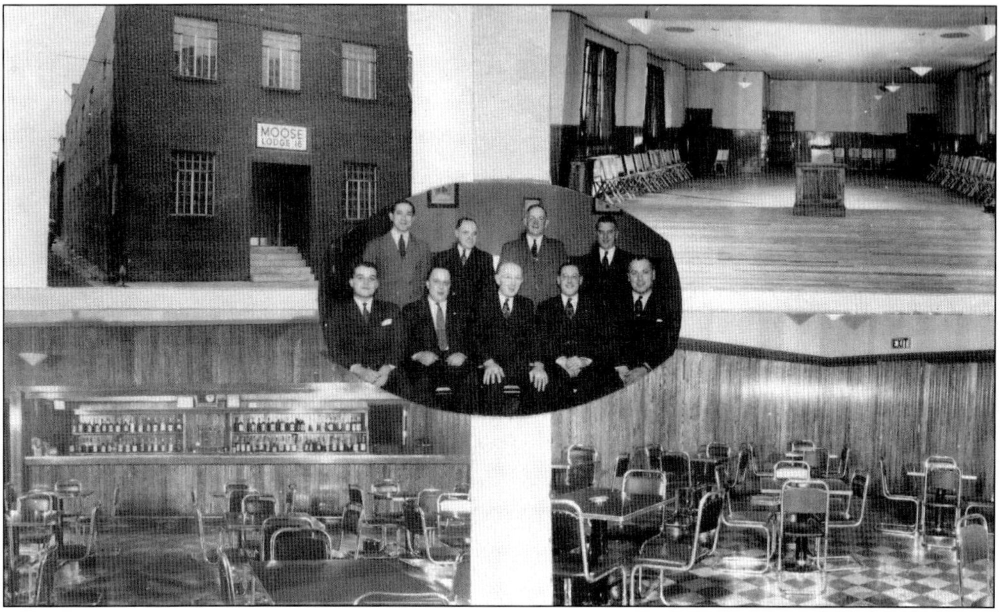

After a disastrous fire, the Loyal Order of the Moose, Lodge No. 16 rebuilt the inside of their meeting and social hall on West Apple Street. This composite photograph from 1948, prepared by photographer Felix Prestia Sr., shows the lodge officers surrounded by interior views of the clubhouse. Fred Munk, the lodge secretary, is seated second from the right. The other officers are unidentified. (Courtesy of Felix Prestia.)

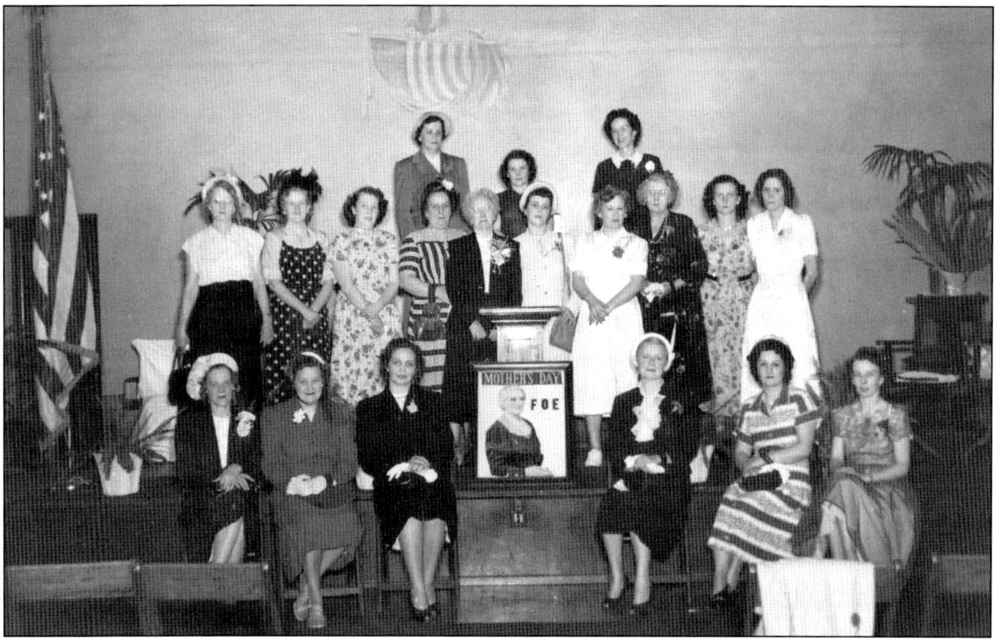

Mother's Day in Connellsville, and elsewhere throughout the region, was commemorated with special events. Seen here is a 1950s celebration at the Connellsville Order of the Eagles. The parent organization of the Eagles supported a movement that led to a 1914 proclamation by Pres. Woodrow Wilson making the second Sunday in May in honor of mothers throughout the nation. (Courtesy of John Ed Dailey.)

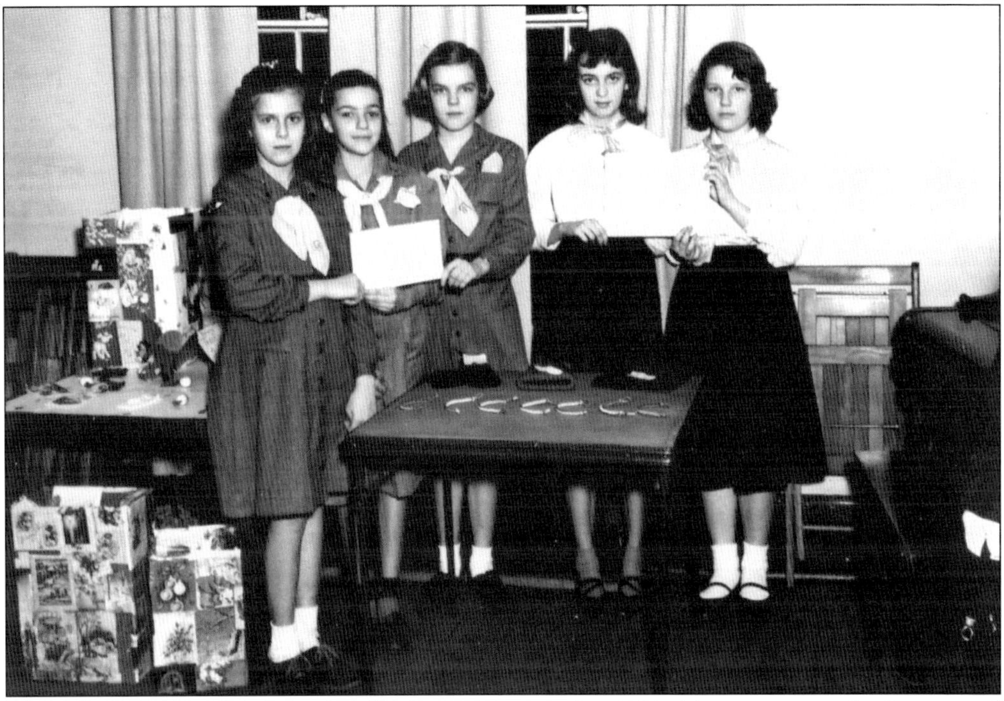

The first Girl Scout troop in Connellsville was organized in 1921. After meeting at the Greenwood Methodist Church, they moved to the Little House on East Crawford Avenue in 1941. This 1950s photograph shows scouts inside the house. They are, from left to right, Jean Lipps King, Mary Ratai Talone, Janice Swallop Jaynes, Nancy Soisson Dye, and ? Schroyer. (Courtesy of the Girl Scout Little House Society.)

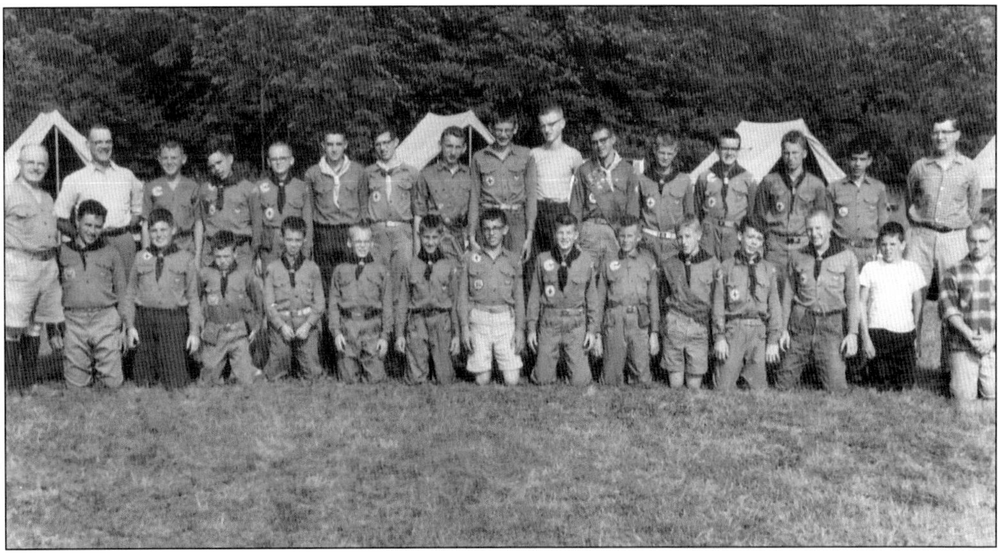

Boy Scout Troop 1 is seen here at Camp Wildwood in 1944. The unit was sponsored by the United Brethren Church and continues to exist as Troop 101, chartered by the Otterbein United Methodist Church. R. Clark Witt, their first Scoutmaster, is on the far left. Witt was also a scout executive for the Fayette County Council and the camp director at Wildwood from 1926 to 1937. (Courtesy of Oscar Tissue.)

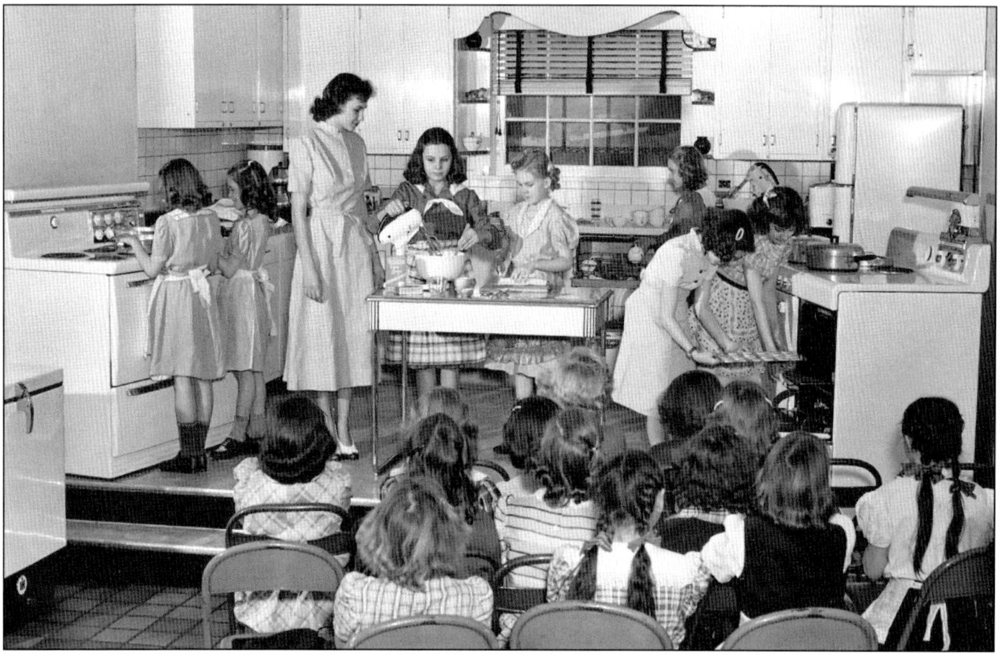

This is what a modern kitchen looked like in the early 1950s. It was located in the basement of the West Penn terminal building. The company used the kitchen to promote and demonstrate the latest electrical appliances. Here, a Connellsville Girl Scout troop is participating in a cooking demonstration. (Courtesy of the Girl Scout Little House Society.)

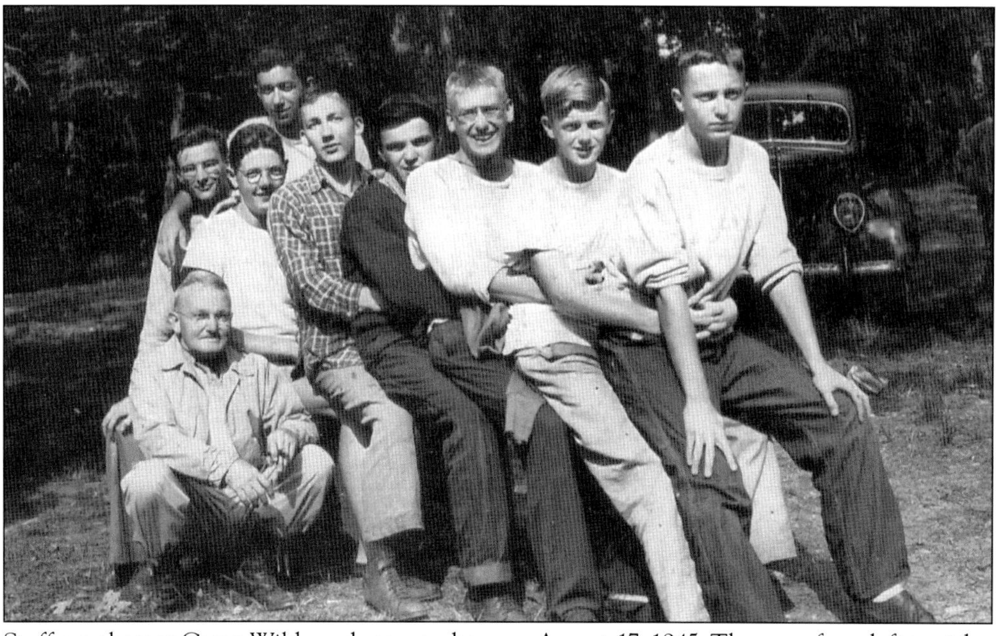

Staff members at Camp Wildwood are seen here on August 17, 1945. They are, from left to right, Earl Graham (adult), Harry Rubin, Dale Newell, Malcolm Crawford, James Dull, Ignassius Soisson, James Wright, Harry Brownfield, and David Barque. Wildwood was located near Normalville, on Mill Run, and served as the Boy Scout camp for the Westmorland-Fayette Council before the opening of Camp Conestoga in 1949. (Courtesy of Oscar Tissue.)

Five

Celebrating and Having Fun

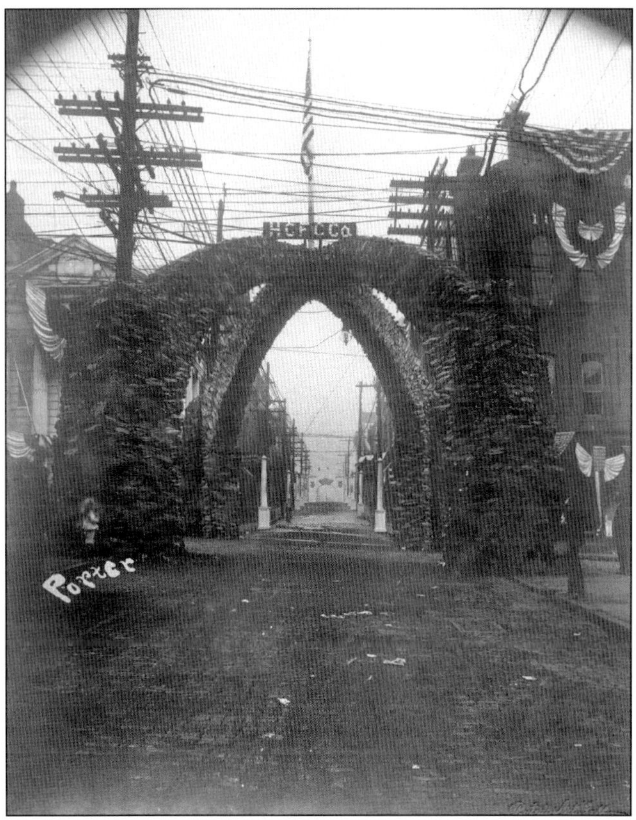

The H.C. Frick Coke Company erected these twin arches over Brimstone Corner for the 1906 Connellsville centennial. They were built from coal and coke and illuminated at night with 657 lights. The coke was manufactured at Trotter, Leisenring No. 1, and Davidson, all of which were owned by Frick. Colonial National Bank is partially visible on the left. The view looks south on Pittsburgh Street.

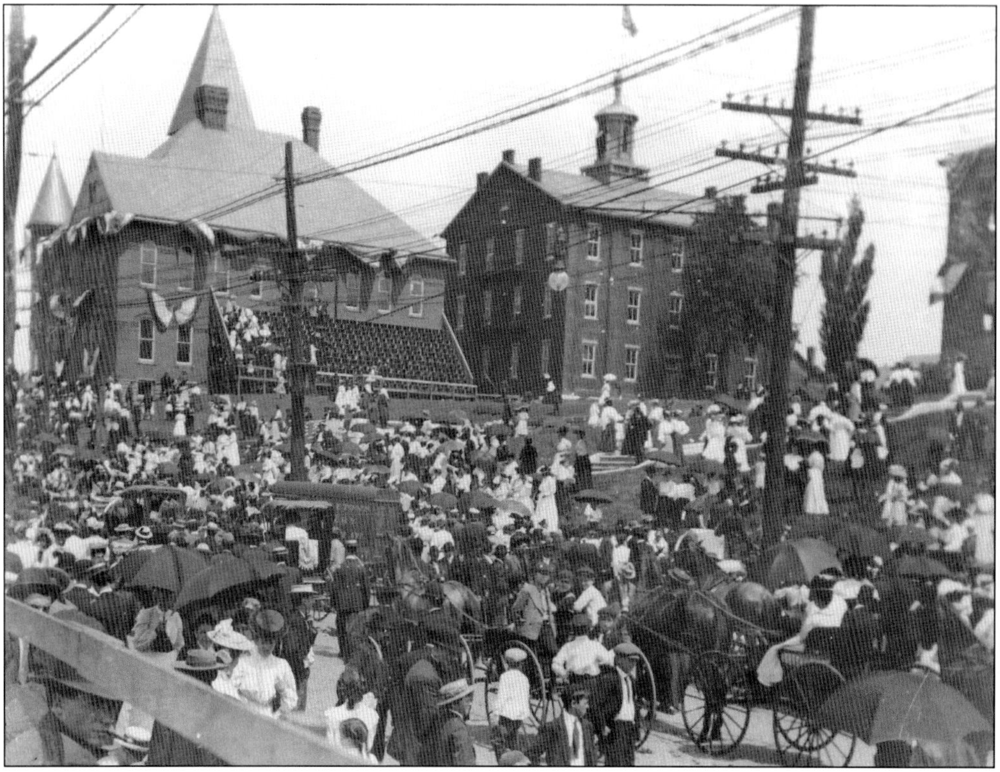

Connellsville residents gather on the lawn of the Carnegie Library during the centennial. The grandstand was constructed for the occasion against the south face of the first high school building. The Union School is the three-story structure on the right. The location served as a venue for speeches, bands, and choral concerts during the four-day celebration held August 14–17, 1906.

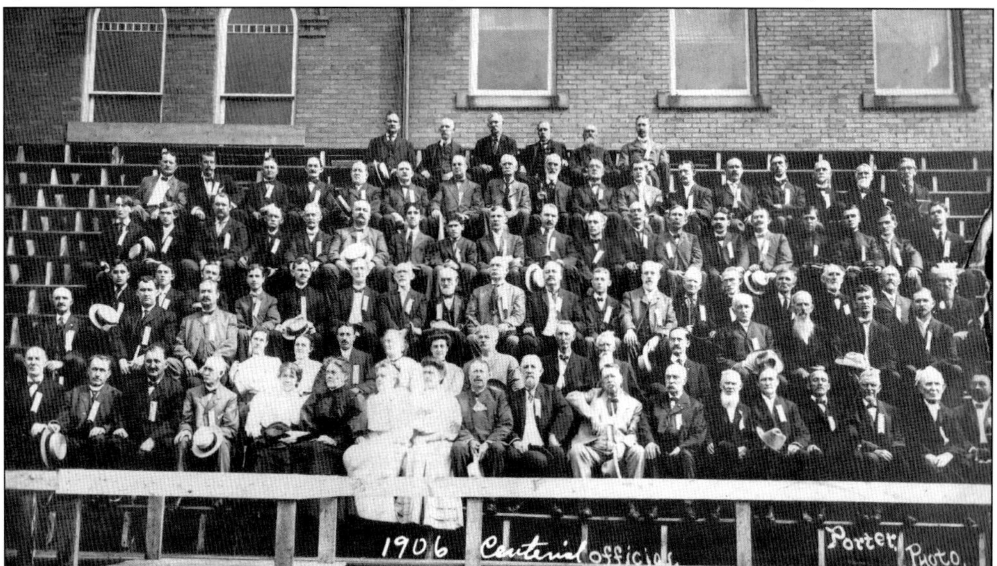

Members of the steering and operating committees that planned the centennial are seen here on the grounds of the library. Rockwell Marietta was president of the association.

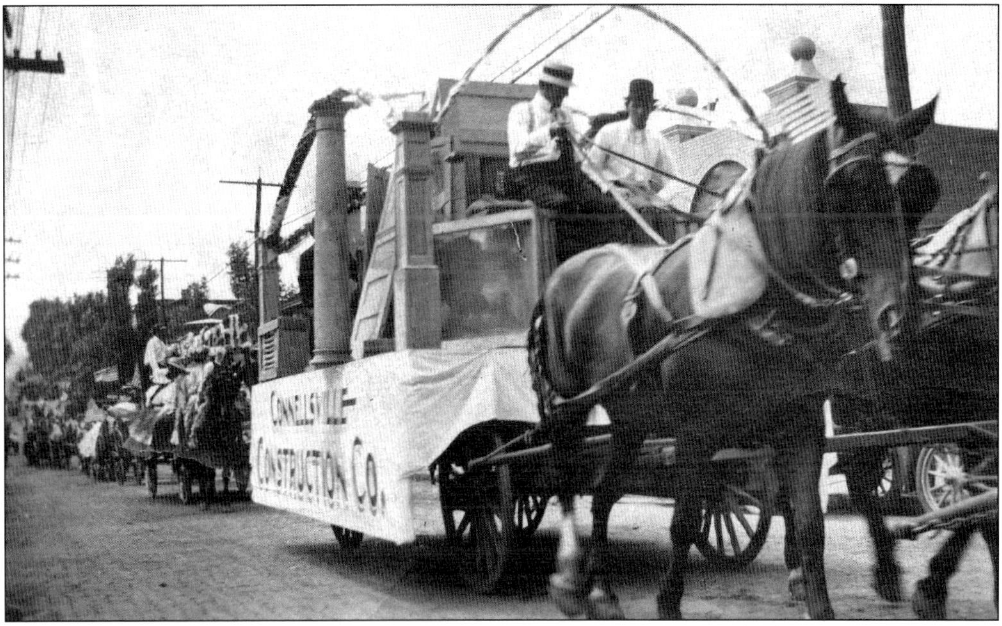

The Connellsville Construction Company had a lot to celebrate with this entry in the 1906 centennial parade. Organized in 1903, it employed nearly 100 carpenters, masons, and laborers. In addition to family houses, the company built the YMCA, Aaron's Furniture Store, the Masonic temple, and the passenger and freight stations for the B&O Railroad.

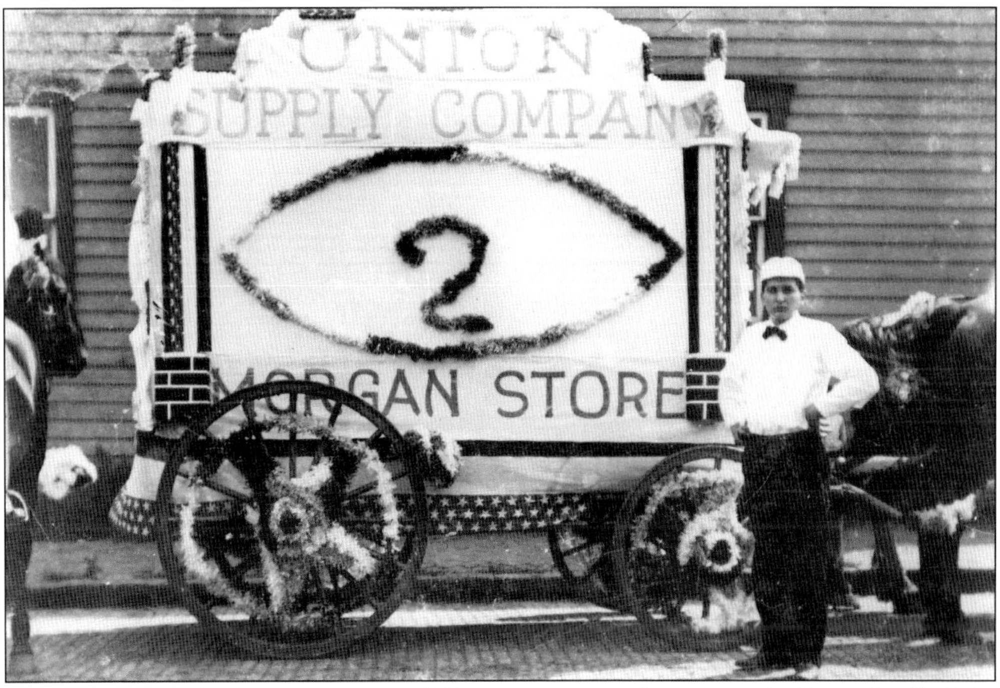

The H.C. Frick Coke Company operated 53 company stores throughout the Connellsville Coke Region under the corporate name of Union Supply Company. During the centennial parade, each store entered a decorated wagon. This is the entry from the Union Supply store at Morgan Station (Broad Ford), which served employees and their families at the Morgan Mine.

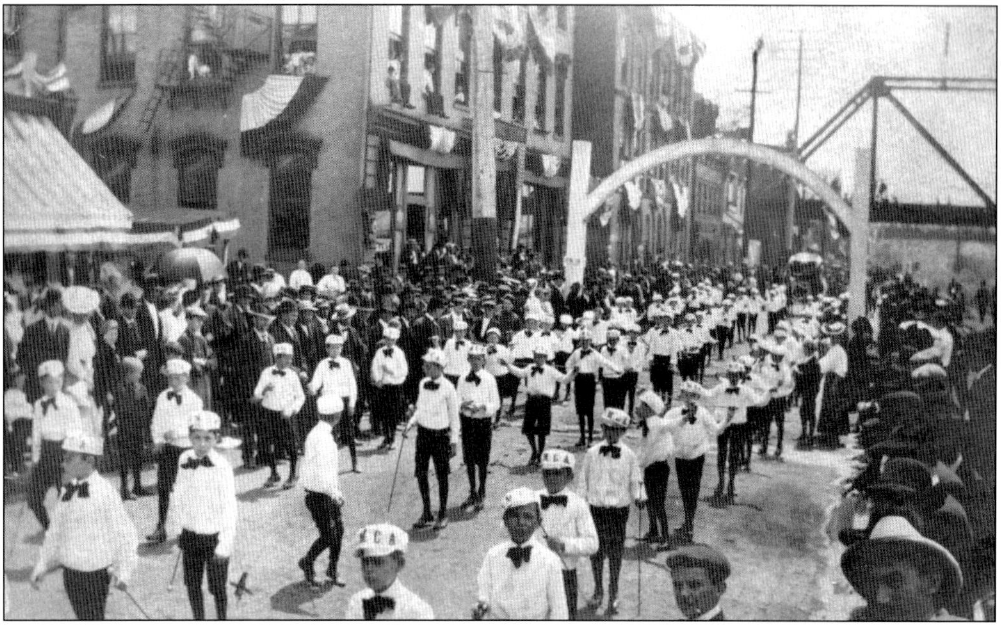

Young YMCA members march on Water Street in Connellsville's centennial parade. Each member wore a white uniform and carried a cane.

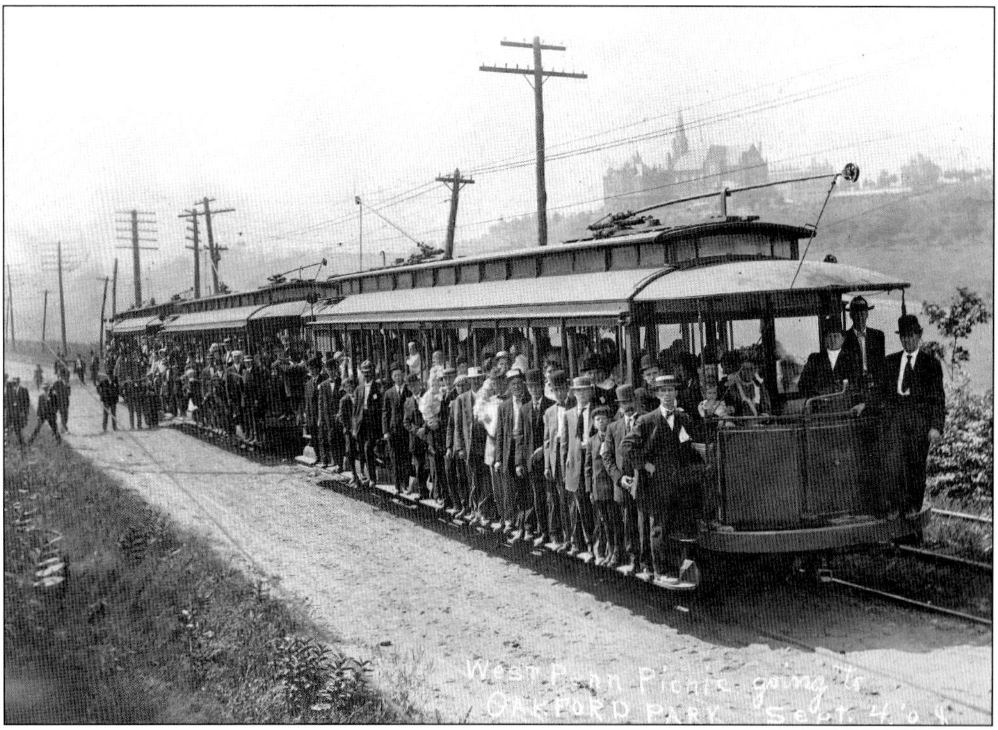

Employees of the West Penn Power Company were traveling to Oakford Park in Jeannette for a picnic when this photograph was taken in 1908. The trolley cars stopped briefly in Greensburg, below Seton Hill College. The school is seen in the upper right of this photograph.

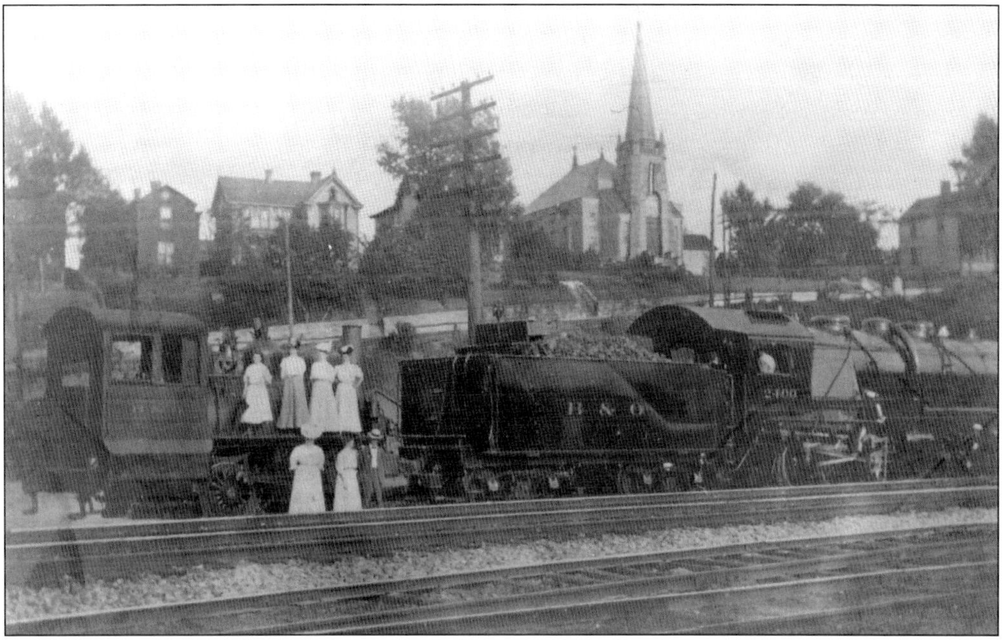

During Connellsville's centennial, the B&O Railroad displayed the world's largest steam engine, called "Old Maude," on a siding near South Arch Street. It attracted these ladies and their escorts for a memorable photograph. Before this celebration, "Old Maude" was featured at the St. Louis World's Fair in 1904.

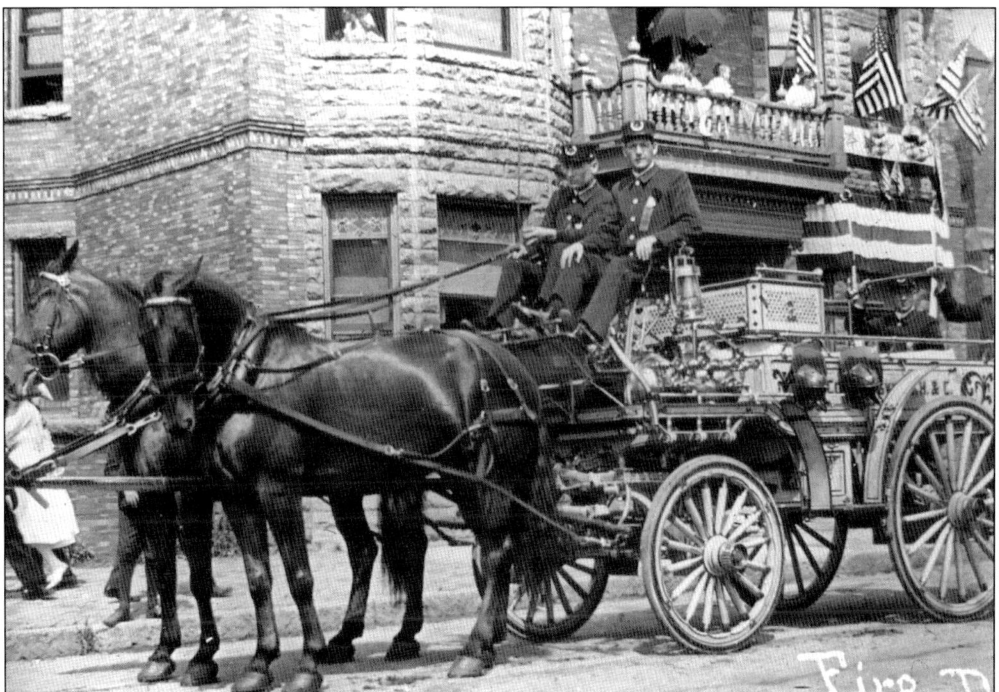

The Connellsville Fire Department entered this horse-drawn equipment in the centennial parade. This photograph was taken in front of the Windsor Apartments on South Pittsburgh Street.

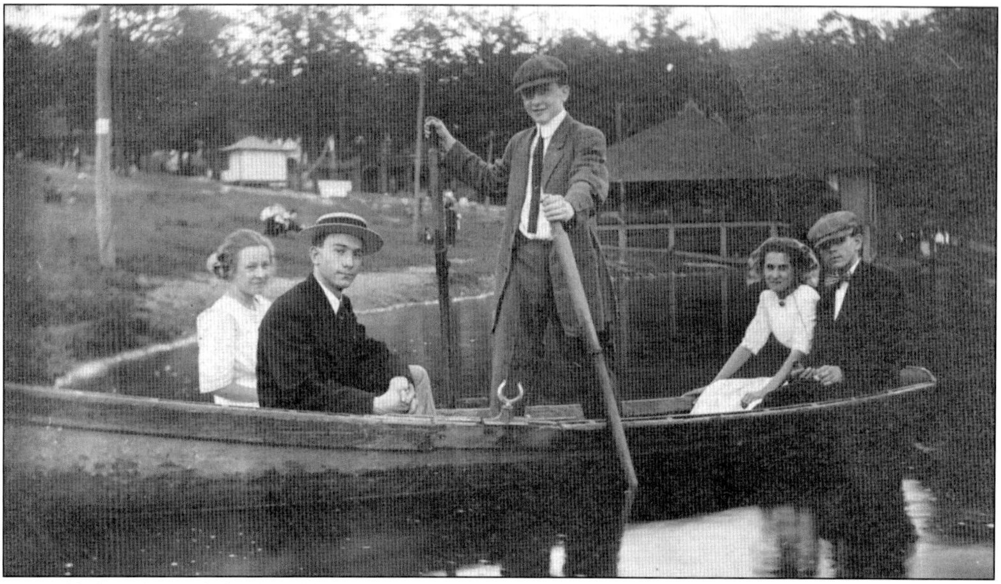

Joe Pallidino (seated, left) and his unidentified friends enjoy boating at Shady Grove Park around 1930. Shady Grove was one of five amusement parks owned by West Penn Railways. They were initially developed in order to increase travel during the evening hours and on weekends. The park, located in Lemont Furnace, is still open to the public and features a large swimming pool and picnic area. (Courtesy of Margaret Molinaro.)

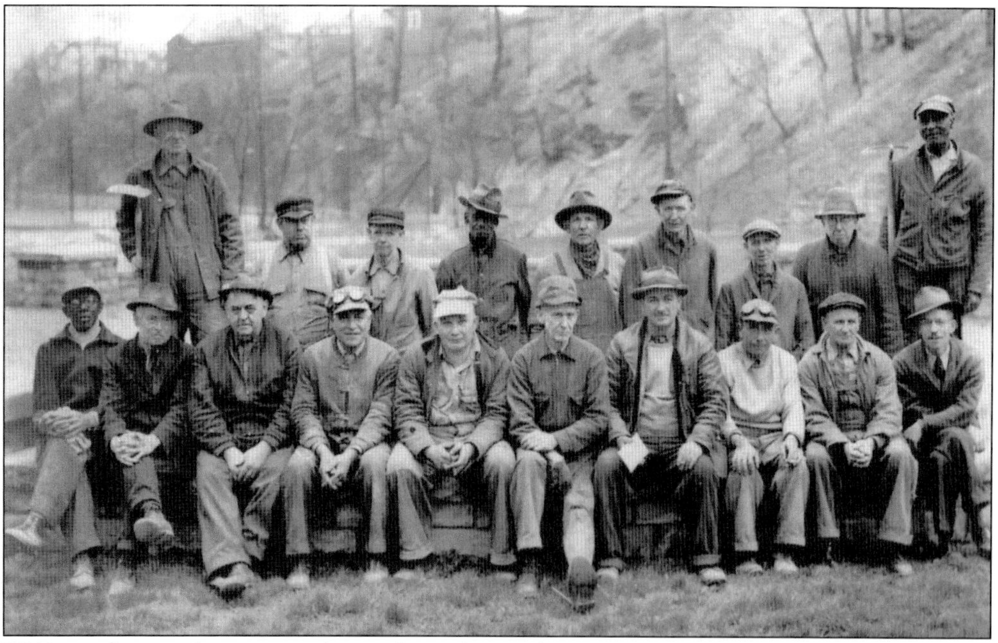

WPA employees who converted an industrial area along Connell Run into East Park assembled for this photograph on April 22, 1943. They are, from left to right, (seated) Thomas Johnson, Victor Baritell, M.J. Cummings, Ralph Emilo, Dewey Smith, A.R. Boyer, William Special, Dan Raymond, George McCune, and R.W. Duckworth; (standing) Frank Martin, E.D. Collins, Joseph Williams, Joseph Webster, Scott Alexander, Steve Urchosko, Joseph White, Luther Long, and Joseph Watts. (Courtesy of Joseph McKitrick.)

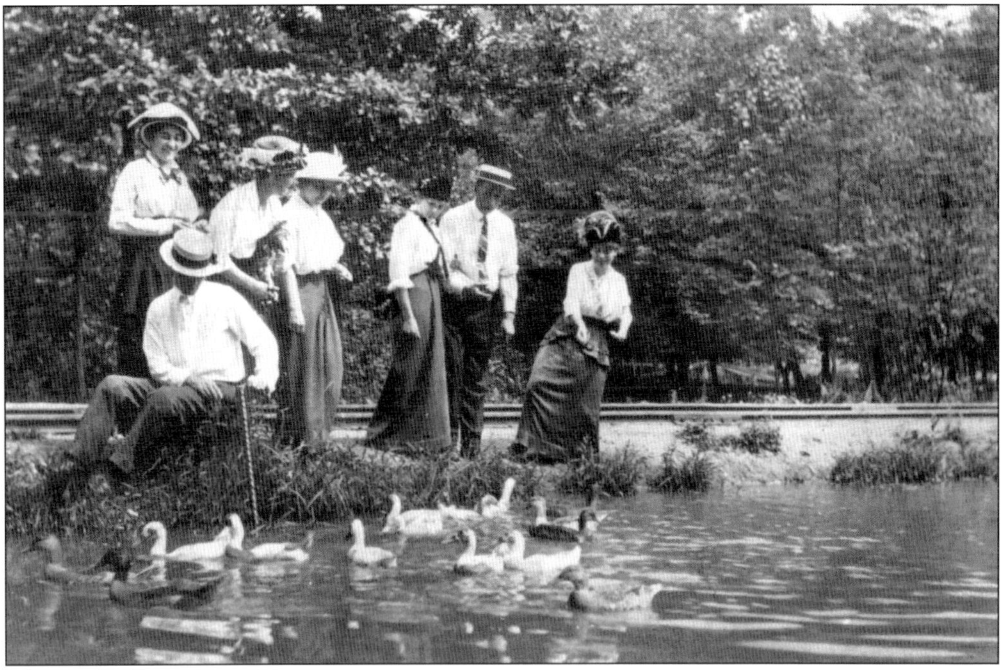

Visitors enjoy the lake at Soisson Park in South Connellsville around 1900. The park was situated above Third Street between Dushane and Park Avenues. It was advertised as a "pleasure retreat" where area residents could experience the outdoors on summer days away from the city. In addition to boating on the lake, the park featured a dance pavilion and refreshment stands. (Courtesy of Scottdale Bank & Trust Company.)

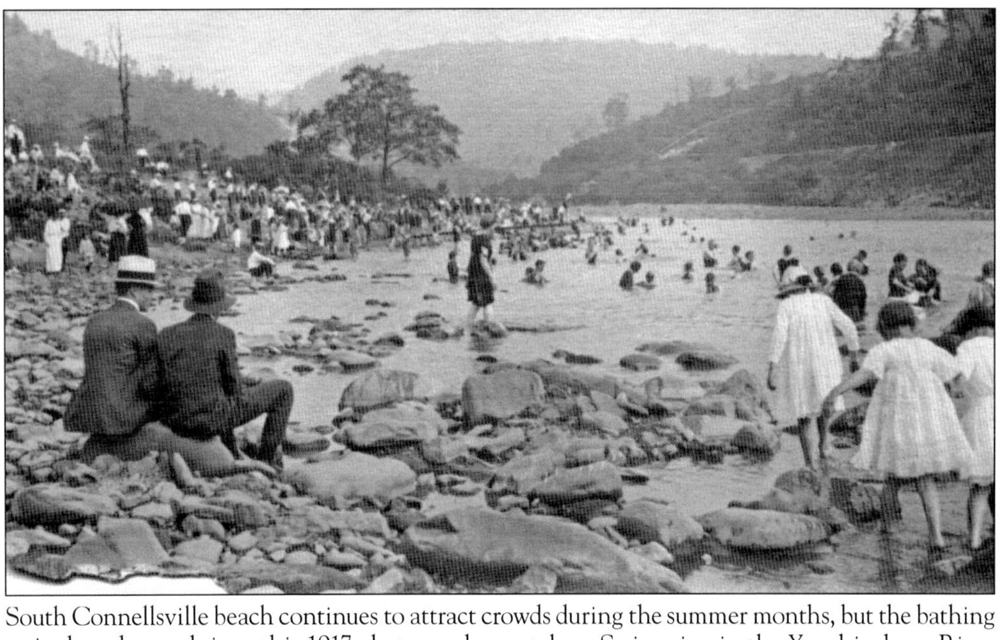

South Connellsville beach continues to attract crowds during the summer months, but the bathing attire has changed since this 1917 photograph was taken. Swimming in the Youghiogheny River always required strength and skill due to the strong current, which unfortunately produced tragic consequences for some who were unfamiliar with its special challenges.

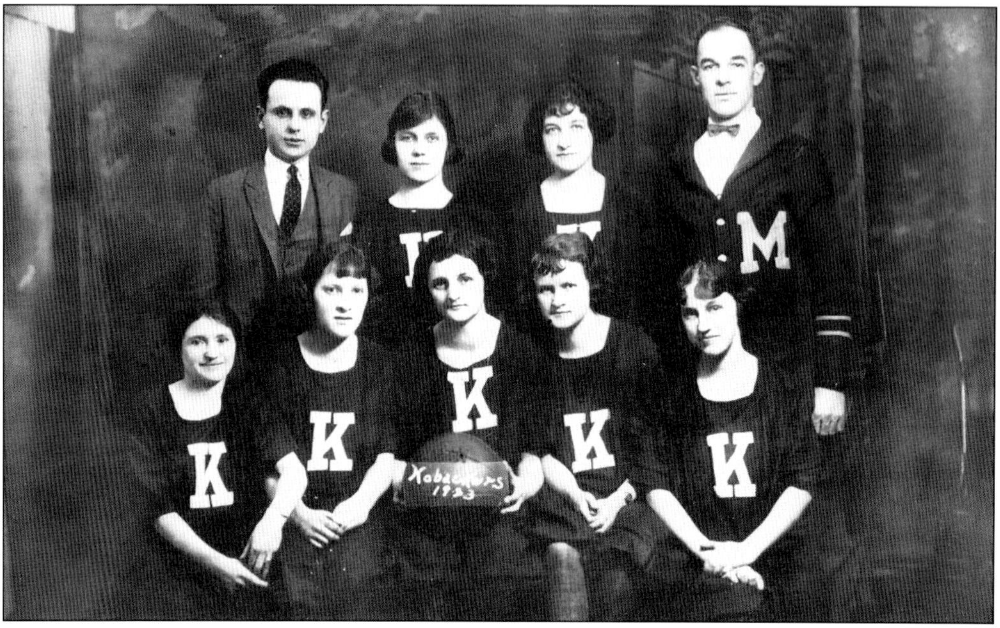

Basketball was a popular sport for women in Connellsville. The Kobacher girls are seen here in 1923. The team was sponsored by Morris Kobacher, who opened the first department store in Connellsville. They were coached by Fred Snell, who is in the second row on the far right. In subsequent years, Snell became a dominant figure in track and field sports throughout the region.

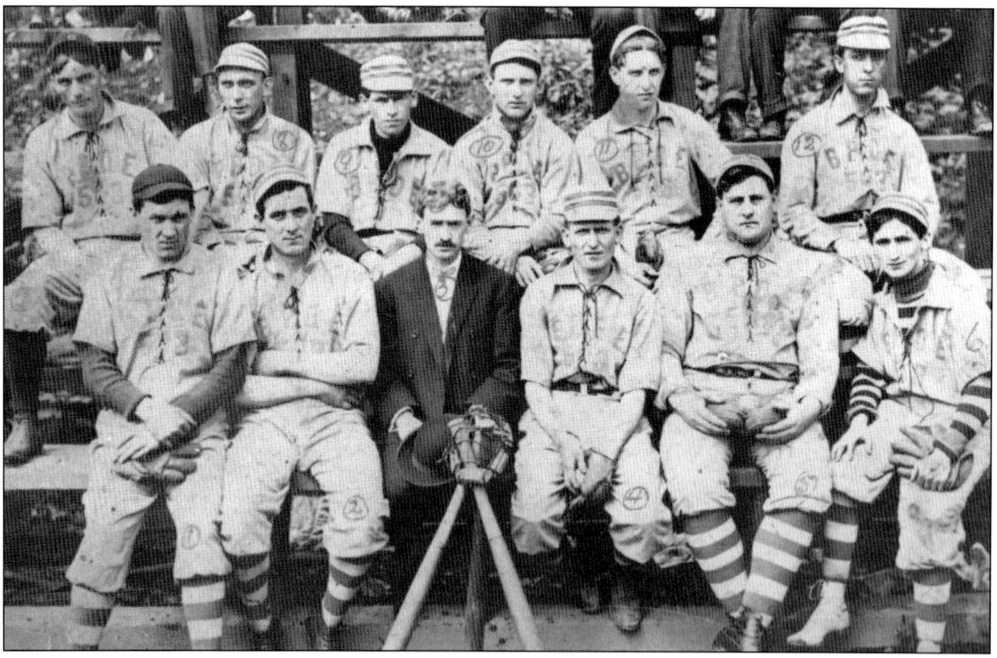

This baseball team was sponsored by Benevolent and Protective Order of Elks No. 503 at Fayette Field in 1918. They are, from left to right, (first row) Harry Irwin, Patrick Irwin, James Mahoney (manager), Jesse Whaley, William Wingenworth, and Bud Noonan; (second row) Lawrence Francis, Dr. John Singer, Raymond Coll, Walter Sneddon, James Purcell, and Guy Swingley. (Courtesy of Paula Salatino Beucher.)

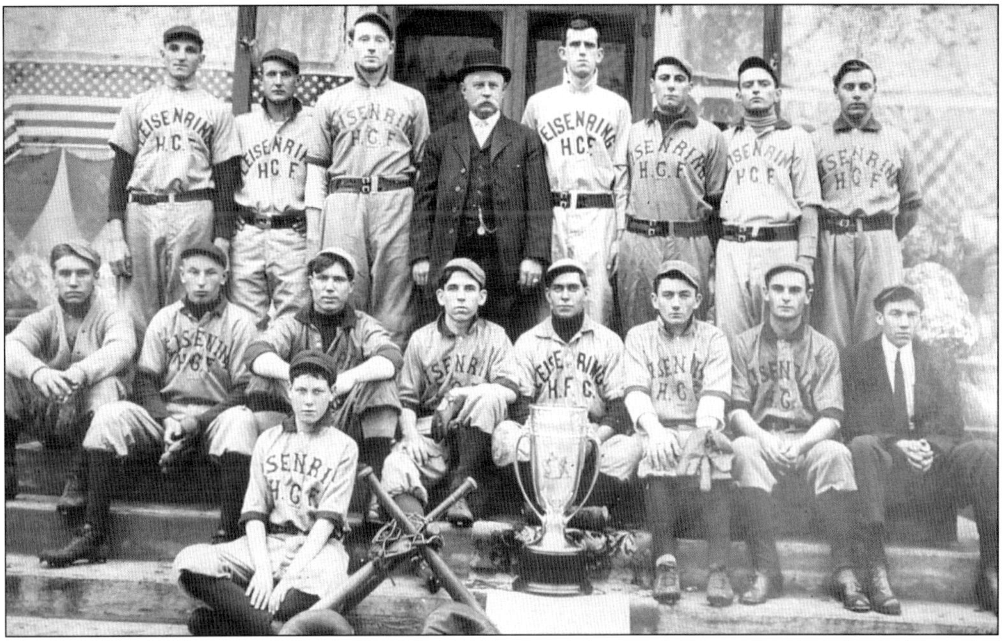

Coal and steel companies sponsored baseball teams that were very popular throughout the Connellsville Coke Region. This photograph from around 1900 shows a championship team from Leisenring. The trophy in the center is probably the Lynch Cup, an award named for Thomas Lynch, who served as general manager and president of the H.C. Frick Coke Company.

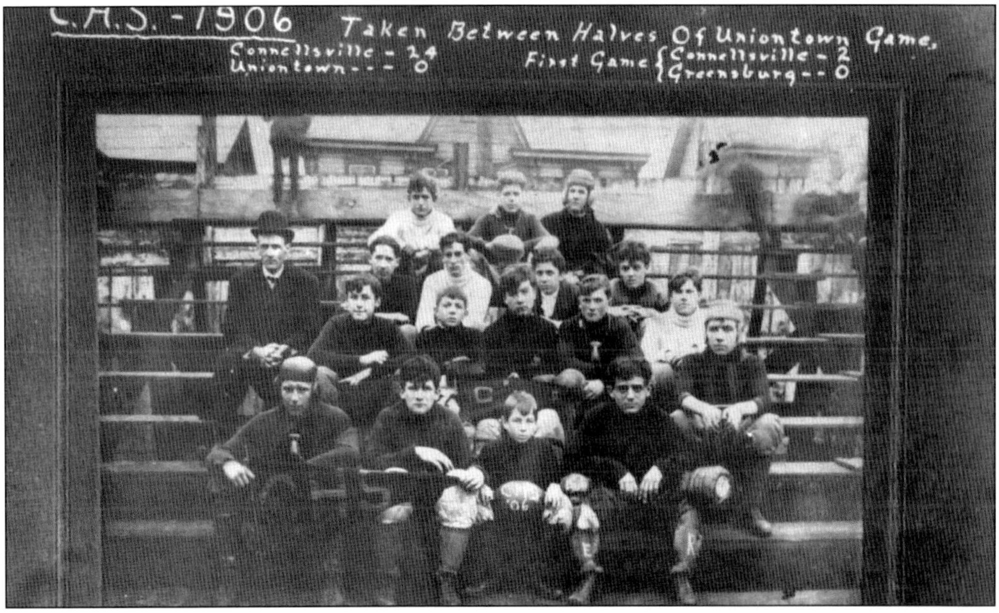

The 1906 Connellsville football team is pictured at Fayette Field with the winning scores. Team members are, from left to right, (first row) Harry Ashe, Ralph Marshall, Jim Munson (mascot), Eli Rosenblum, and Bob White; (second row) Max Scott, Bob Morton, Phil Swartzwelder, Minnie Casper, and Eugene Bishop; (third row) W.S. Deffenbaugh (school superintendent), Marshall Mulford, "Button" Munk, Edgar Hicks, and Bill Brickman; (fourth row) Herman Rosenblum, "Cotton" McCormick, and John King. (Courtesy of John Ed Dailey.)

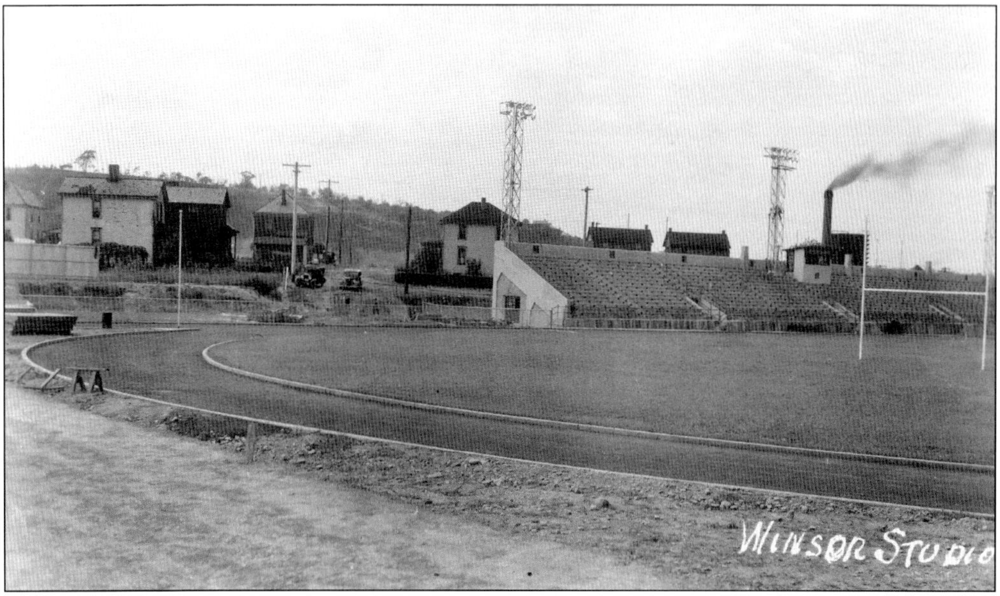

Campbell Stadium, another Works Progress Administration project in Connellsville, was regarded as one of the best stadiums in Western Pennsylvania. Rooms for the home team and the visitors were equipped with showers and restrooms. The first game was played on September 9, 1938. John Woodruff was a special guest; he and some of his teammates from the University of Pittsburgh sprinted around the track during the opening.

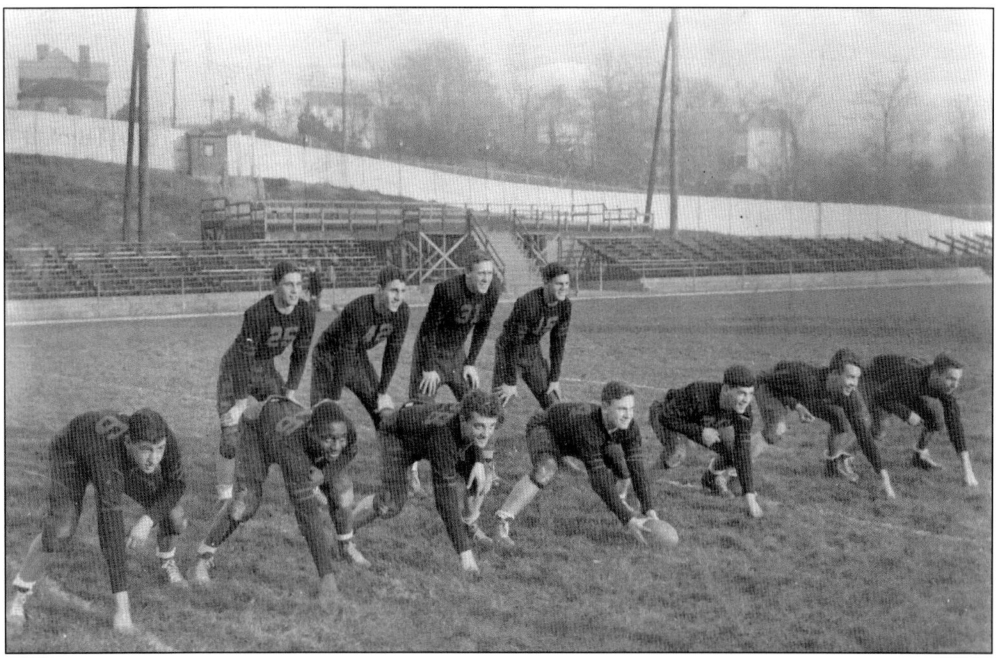

The 1941 offensive lineup for the Cokers set an enduring standard for excellence in the game and in lifetime achievements that is unequaled in the annals of competitive sports in Connellsville. The players are, from left to right (first row) Richard Dixon, Albert Blannon, Louis Scacchi, Harold Stefl, Gary Feniello, Jim Smyth, and B. Olsweski; (second row) John Lujack, John "Wally" Schroyer, Alfred Bieshada, and David Hart. (Courtesy of John "Wally" Schroyer.)

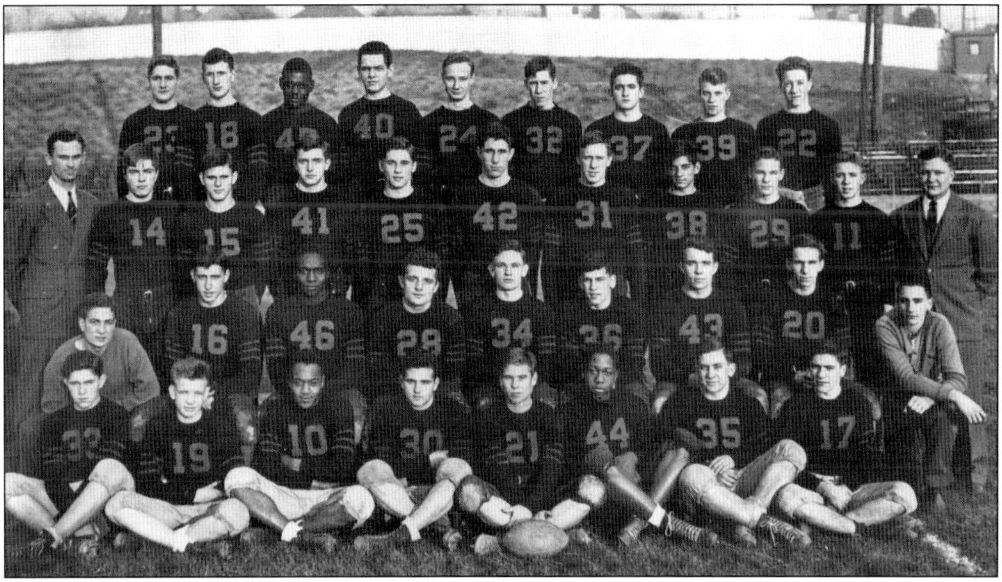

This photograph shows the entire 1941 Connellsville High School football team, which was undefeated in eight games. In the last game, it tied with Brownsville. It did not compete in the playoffs due to an insufficient number of Gardner points. The team was coached by Art Ruff, who stands on the far right in the third row, and Ed Spotts on the far left. (Courtesy John "Wally" Schroyer.)

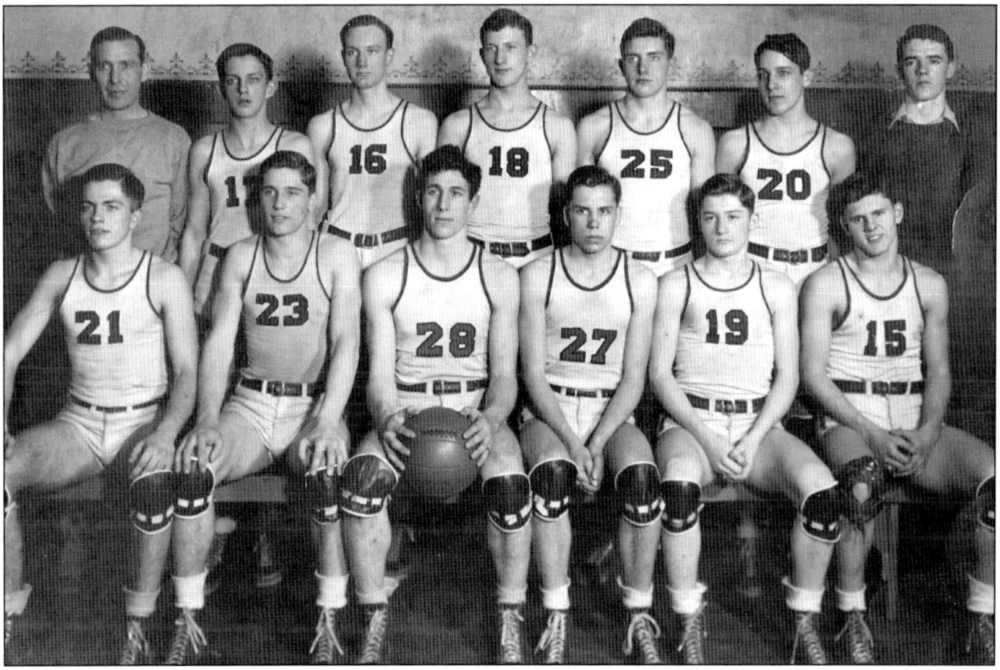

The 1941–1942 Cokers basketball team included, from left to right, (first row) John Richards, John Lujack, John "Wally" Schroyer, Chris Siesky, Robert Randolf, and Dave Hart; (second row) William Dolde (coach), Howard Schrum, Richard Dilworth, Paul Litavic, Richard Pitzer, Ronald Daberro, and James Wolfe. Of the 13 members, 7 also played on the famed 1941 football team. (Courtesy of John "Wally" Schoyer.)

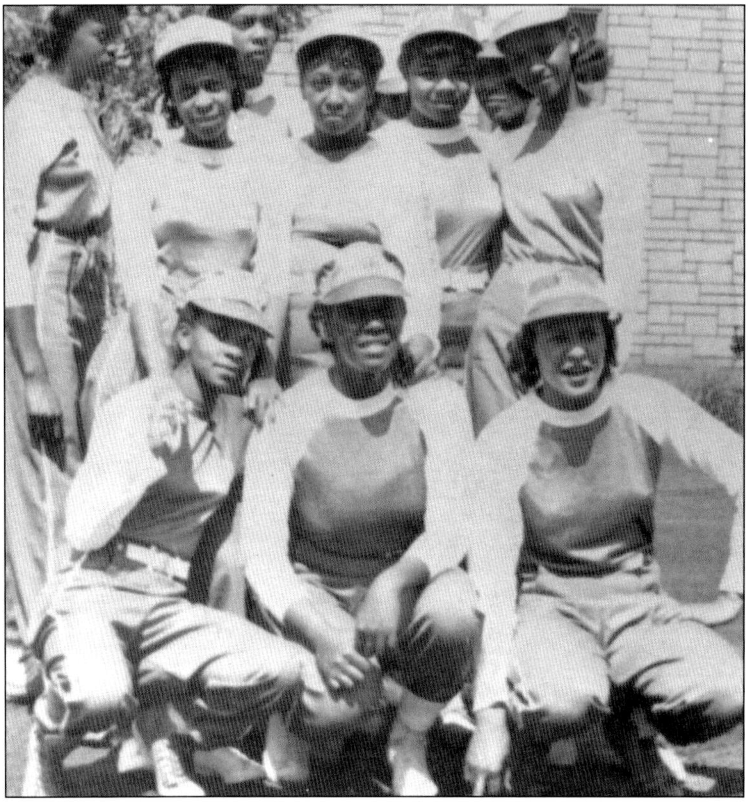

The Trotterettes softball team gathers before a game in 1941. They are, from left to right, (first row) Eleanor Ricks, Betty Edwards, and Nannie Robinson: (second row) Mamie Ricks, Augie Parrish, and Lorraine Davis: (third row) Shirley Copeland, Gwen Braxton, Patty Hunter, and Kathleen Thompson. (Courtesy of Ted Davis.)

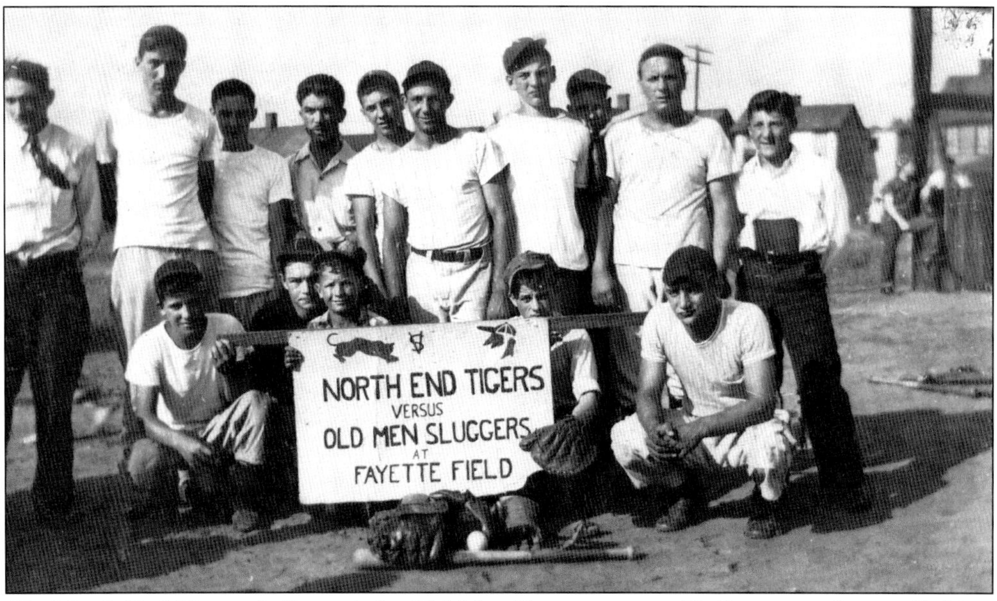

The North End Tigers challenged the Old Men Sluggers to a softball game at Fayette Field in 1944. The team members seen here are, from left to right, (first row) Bob Galasso, Jack Lockery, "Chappie" Hartz, "Juju" Baldi, and Angelo Alisanrino; (second row) Jim Quinn, Tony Pitchman, Andy John, John Cappa, Pete Hartz, Lou Molinaro, unidentified, Bill Cowan, "Gutsy" Hartz, and Henry Molinaro. (Courtesy of Felix Prestia.)

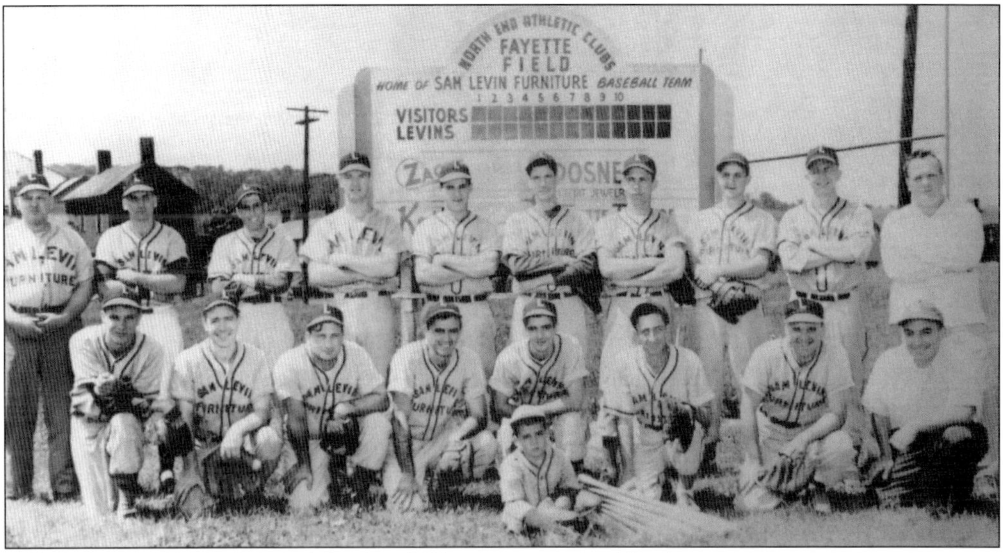

Fayette Field was the home of this baseball team, sponsored by the Levin Furniture Company in 1955. The players are, from left to right (first row) Harold "Snipe" Hartz, Jack Kurtz, John Caringalo, Chris Wagner, Paul Trafecanty, George Kearns, Fred "Buzz" Barnhart, and Norm Bernstein; (second row) "Hutch" McGinnis, Jack Marshall, unidentified, Herb Kurtz, Bob Glasso, Bill Kurtz, Lester Cox, "Bobo" Welsh, Larry Condiff, and "Gutsy" Hartz. (Courtesy of Howard Barnhart.)

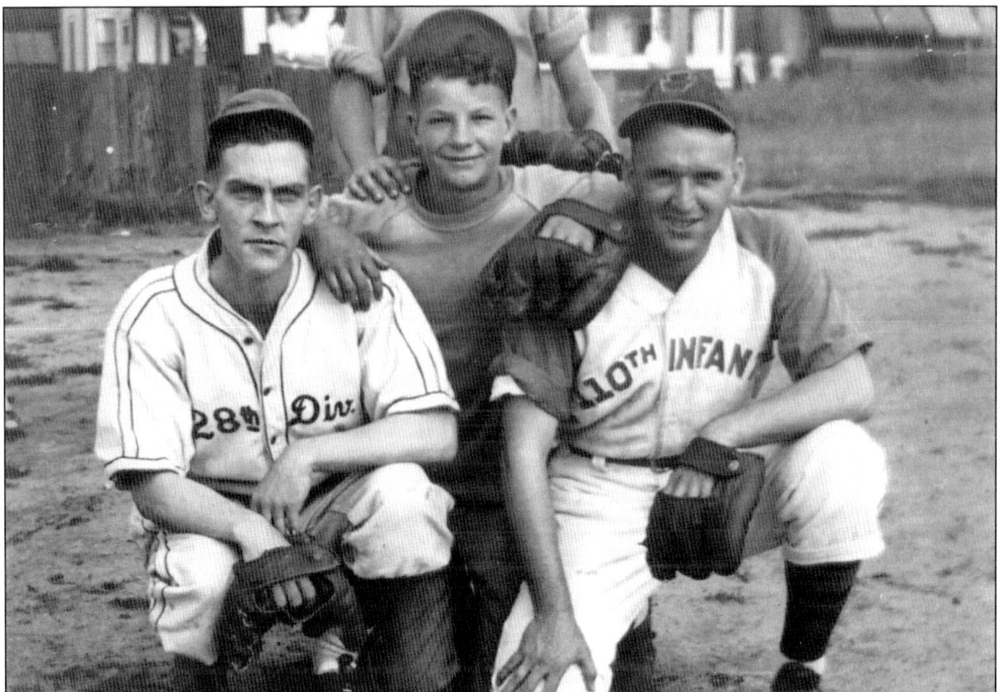

Many veterans were members of sandlot teams that played baseball at Fayette Field. L. Wendell Smith (left) and Fred "Buzz" Barnhart (right) were two of these players. They are dressed in the jerseys of their military units; they played for the 28th Division and the 110th Infantry baseball teams. Charles "Chappy" Hartz is the batboy between the two veterans. (Courtesy of Howard Barnhart.)

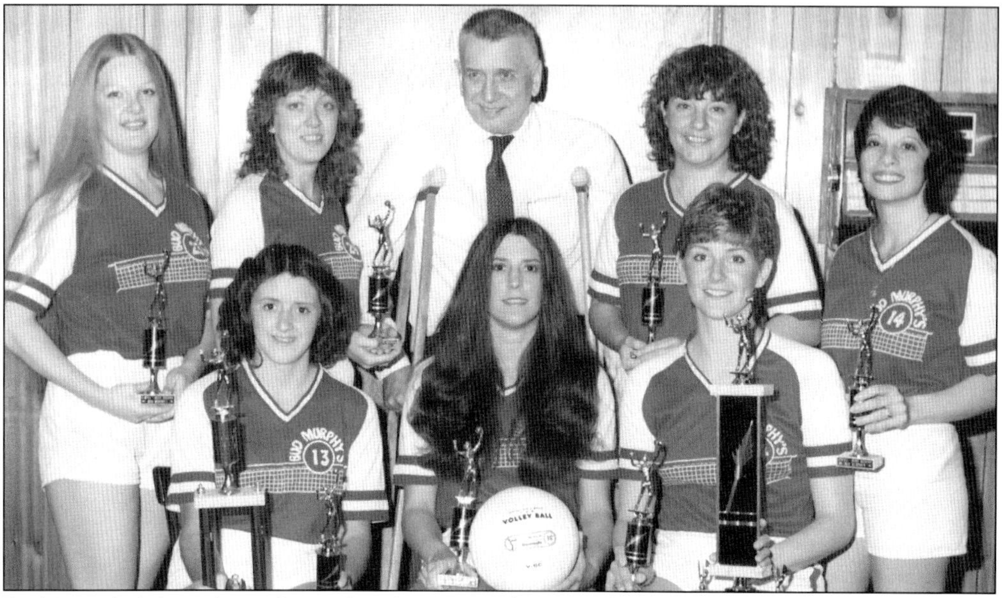

Bud Murphy's has always been the place to go for pizza, but Murphy's real legacy was sports, specifically the teams his family sponsored for nearly 60 years. A championship volleyball team is seen here in the late 1970s. The players are, from left to right, (first row) Helen Wisolowski, Cheryl Welc, and "Indy" Williams; (second row) Melaine Keller, Kathy Hillen, Bud Murphy, and two unidentified girls. (Courtesy of Bud Murphy's.)

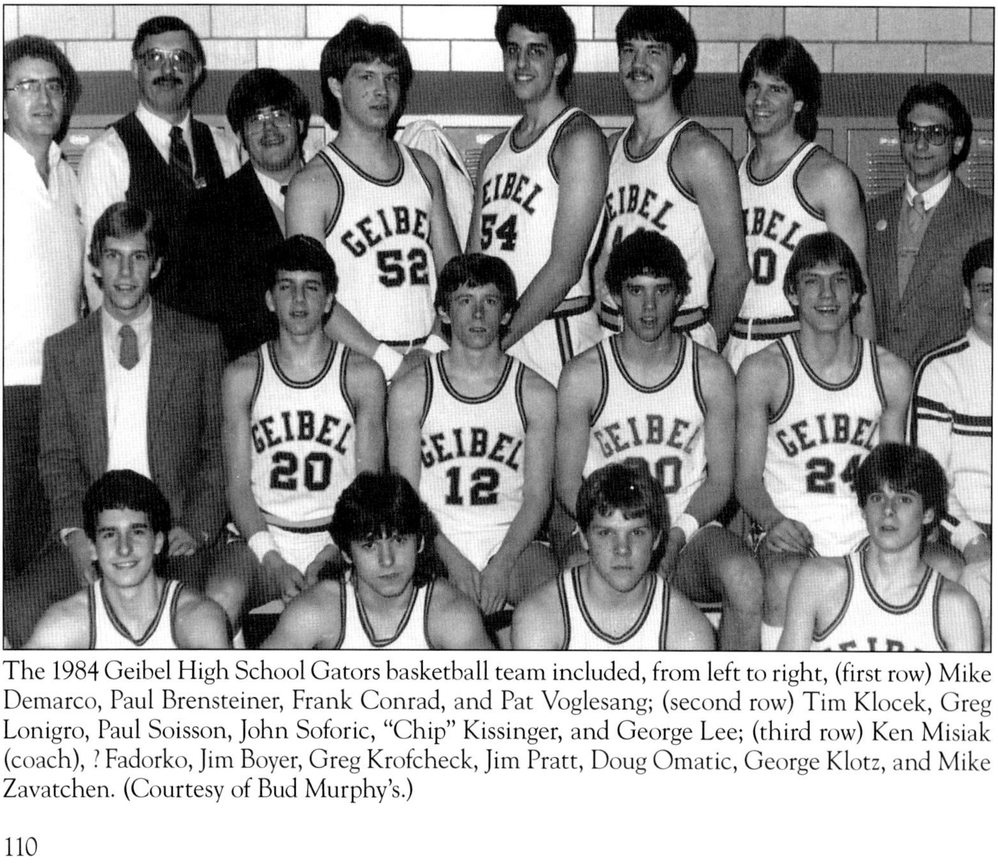

The 1984 Geibel High School Gators basketball team included, from left to right, (first row) Mike Demarco, Paul Brensteiner, Frank Conrad, and Pat Voglesang; (second row) Tim Klocek, Greg Lonigro, Paul Soisson, John Soforic, "Chip" Kissinger, and George Lee; (third row) Ken Misiak (coach), ? Fadorko, Jim Boyer, Greg Krofcheck, Jim Pratt, Doug Omatic, George Klotz, and Mike Zavatchen. (Courtesy of Bud Murphy's.)

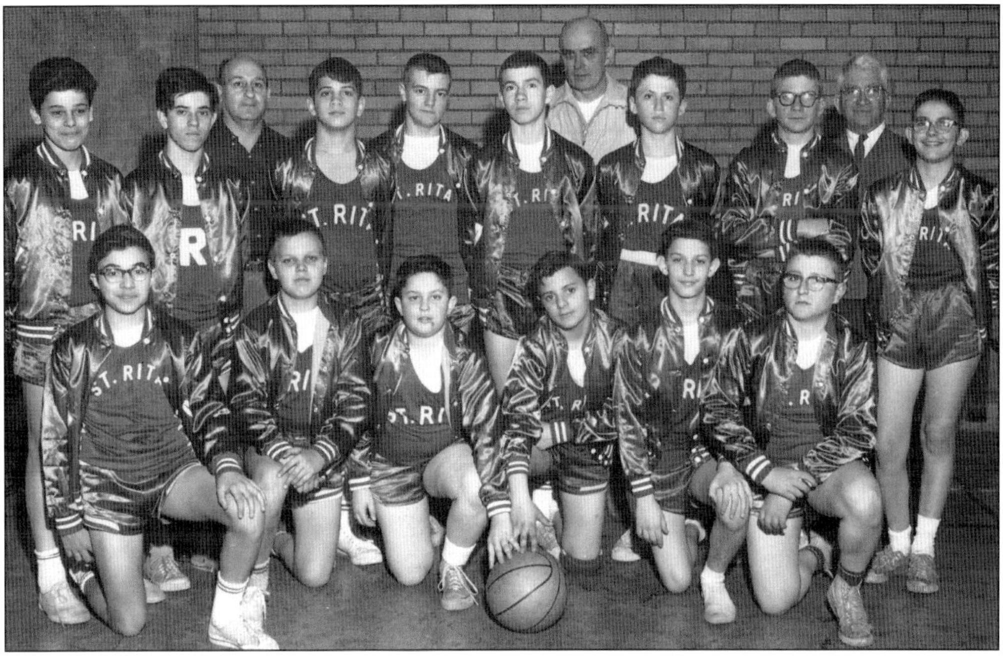

St. Rita's sponsored a youth basketball team for 60 years. Seen here are some league members in 1965. They are, from left to right (first row), Gerry Delligetti, William Sapanaro, Jeff Osler, Ronald Fetsko, Allen Renzi, and Joseph Rovesscchi; (second row) Francis Cricco, Angelo Capo, Pete Salatino (coach), Robert Cavalier, George Cofert, Bob Turek, James Magarrity (coach), Frank Dunaway, Mike Quinn, Sam Natale (league director), and Charles Rosendale. (Courtesy of Paula Salatino Beucher.)

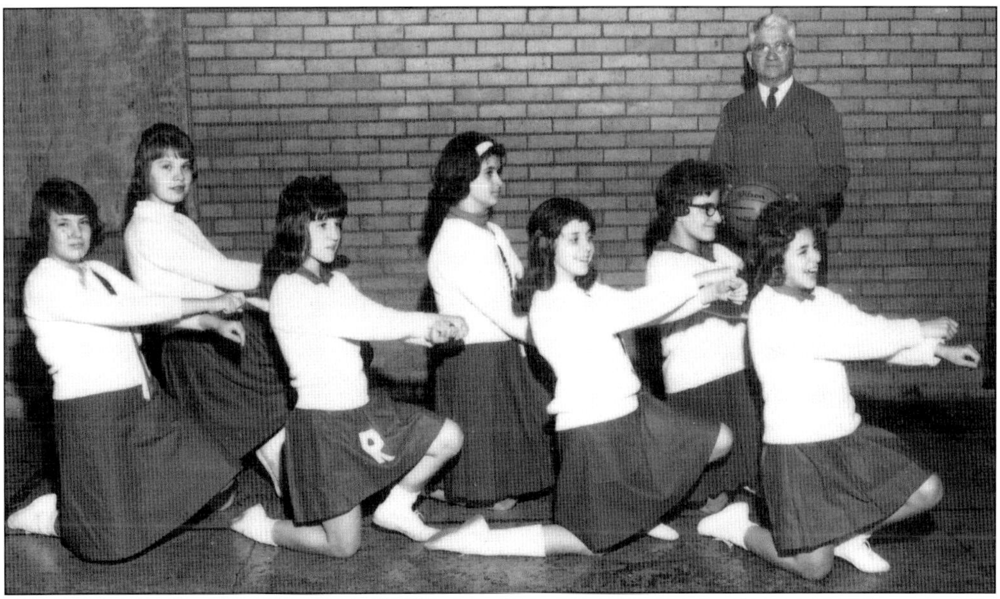

Cheerleaders for the St. Rita's basketball league demonstrate a routine during a 1965 game at the Crawford School. They are, from left to right, Pat Sapanaro, Barbara Bernardo, Anita Gross, Cheryl Orbin, Yvonne Rulli, Cathy Kooser, and Yvonne Trafecanty. Ralph Natali is holding a game ball. (Courtesy of Paula Salatino Beucher.)

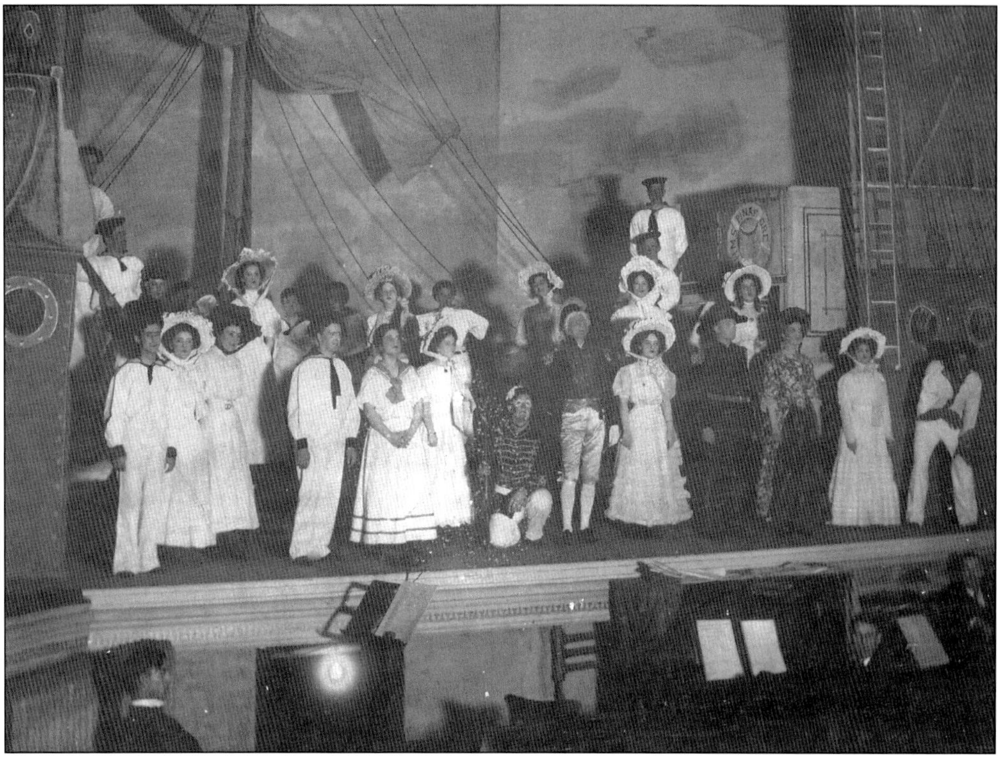

This theatrical performance of *H.M.S. Pinafore* took place at the Newmyer Opera House around 1910. Some of the cast members included Betty Burke, Loma Cole, Margaret Newcomer, Irene Miller, Mildred Miller, Leona Collins, Maria Gemas, Mary Gobright Irene Holland, Hazel Colvin, John Davis, Joe Gobright, and Harry Sheets. The opera house, constructed by Porter Newmyer, was on the corner of North Pittsburgh and Peach Streets.

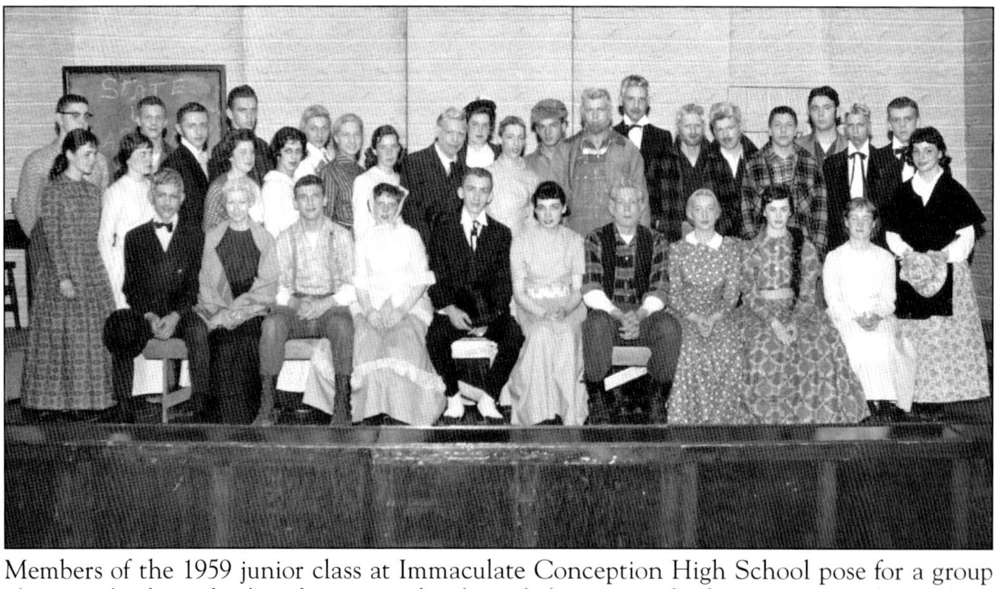

Members of the 1959 junior class at Immaculate Conception High School pose for a group photograph after a final performance of a play titled *Hoosier Schoolmaster*, written by William Spence. All 48 class members were featured in the production.

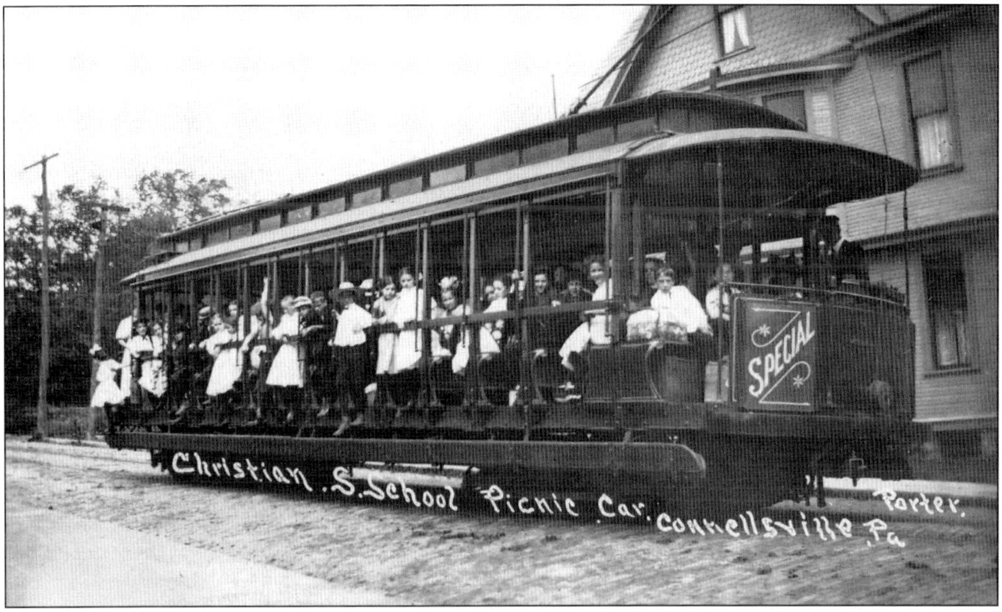

Trains and trolleys provided mobility for work and leisure before the widespread use of automobiles. This 1890s photograph shows a youth group from Connellsville Christian Church traveling to a picnic aboard a West Penn summer car. The photograph was taken at Leisenring. (Courtesy of Bert Swartzwalder.)

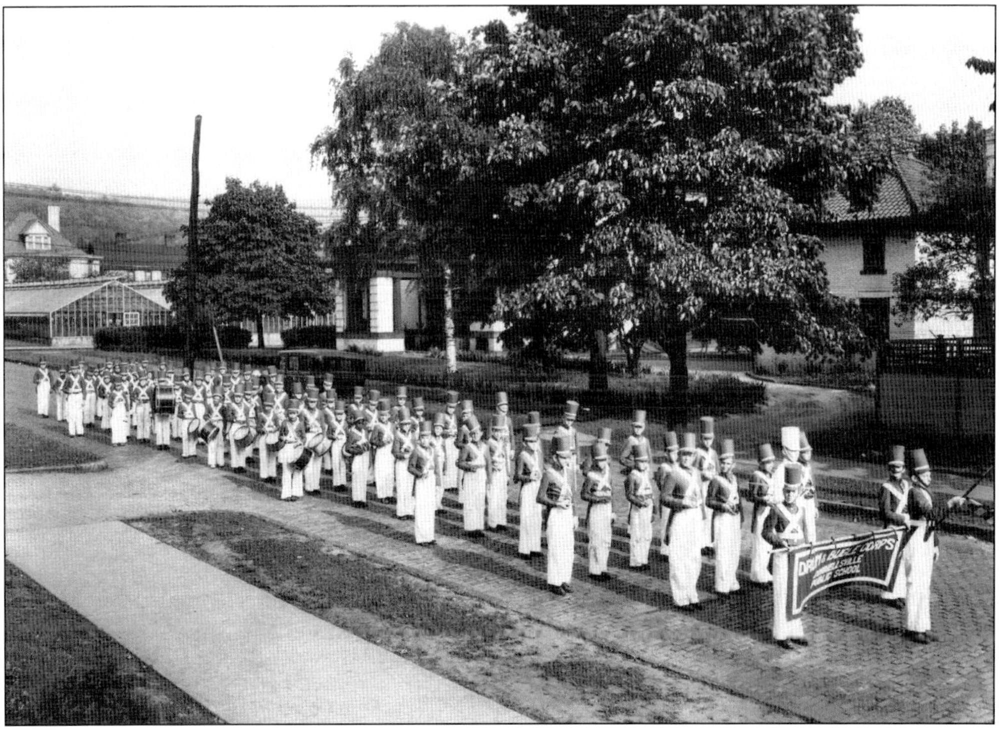

The Connellsville Public School Drum & Bugle Corps assembled for this photograph on West Washington Avenue, near the National Guard armory, in the 1930s. The De Muth Greenhouse is visible in the left background.

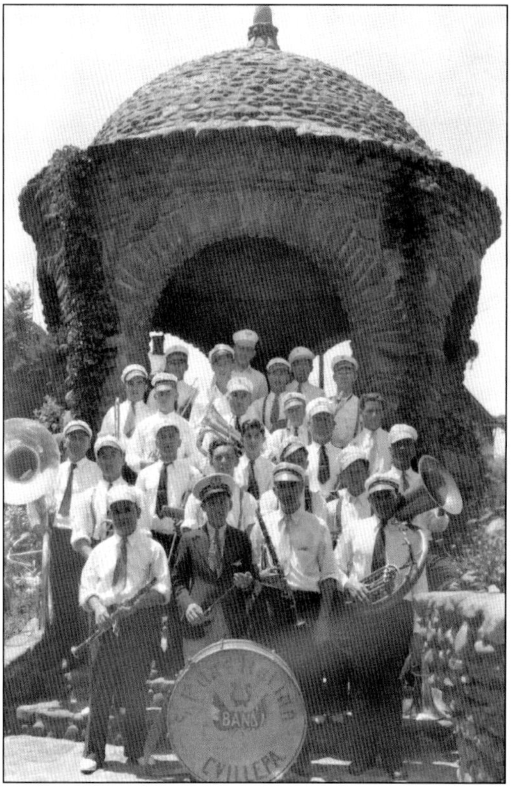

The Molinaro Band is seen at left in 1937, when it was called the St. Rita's Italian Band. The group was originally known as the Royal Italian Band. Michael Molinaro, standing behind the drum in a dark coat, organized the association in 1913. When the baton was passed to his brother Amedeo, the name was changed to the Molinaro Band. During World War II, the band greeted troop trains at the B&O passenger depot with patriotic music. The New Haven Hose Company Band is seen below in the 1950s. A large number of its members also belonged to the Molinaro Band. Its directors included Amedeo Molinaro, Carmine Molinaro Sr., and Lloyd Butts. (Below, courtesy of New Haven Hose Company.)

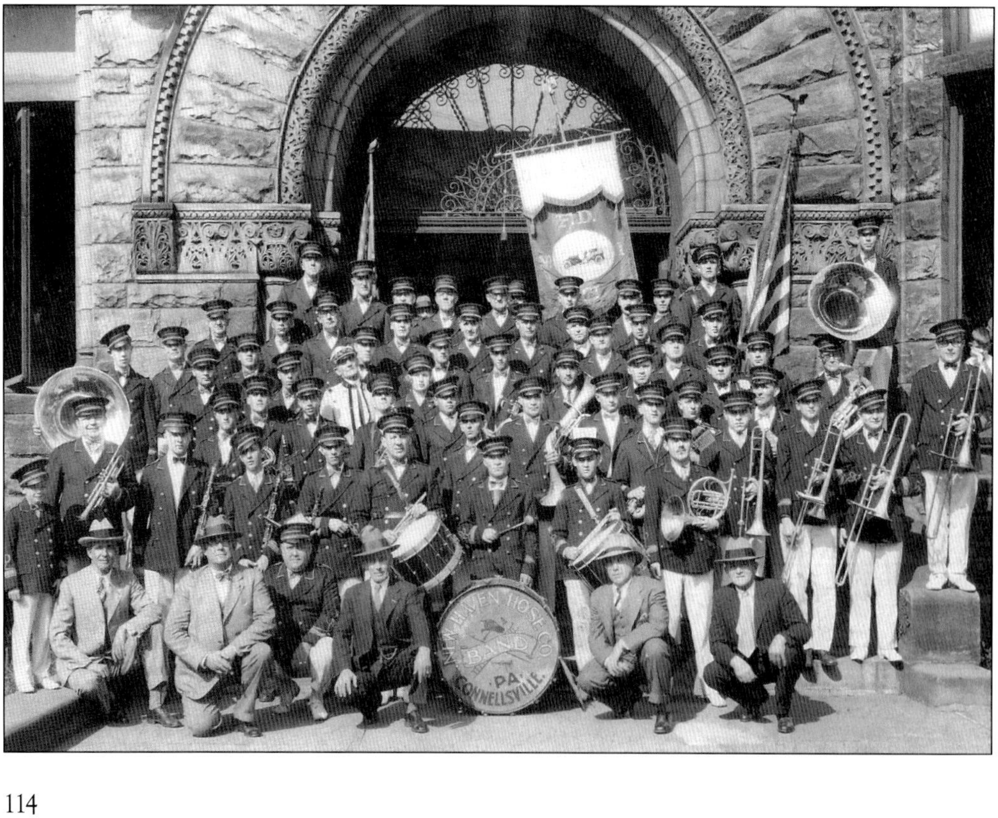

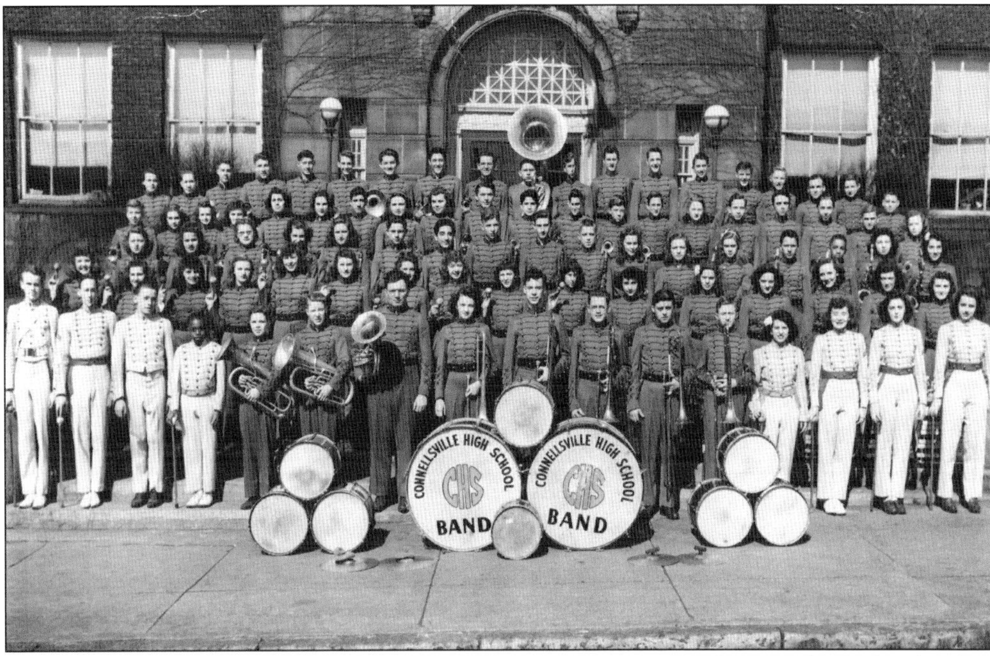

The Connellsville High School band is seen here in the 1950s in front of the high school on East Fairview Avenue. The band, one of the largest in the region, was easily recognizable in its orange and black uniforms.

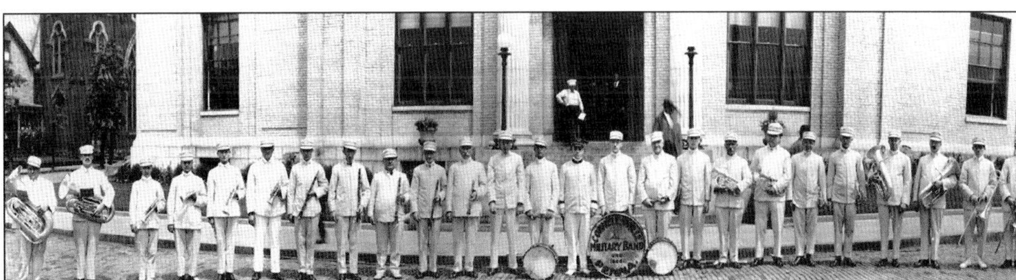

The Connellsville Military Band stopped in front of the Arch Street post office for this photograph around 1920. The band was organized in 1886. During the four-day Connellsville centennial celebration, the band had a prominent role. At the opening ceremony on Tuesday, August 14, 1906, the military band accompanied a grand chorus of several hundred schoolchildren who sang patriotic songs. On two subsequent days, August 15 and 16, it led parades, marching directly behind the color guards. It was estimated that the second parade, sponsored by Connellsville merchants, was over one mile in length.

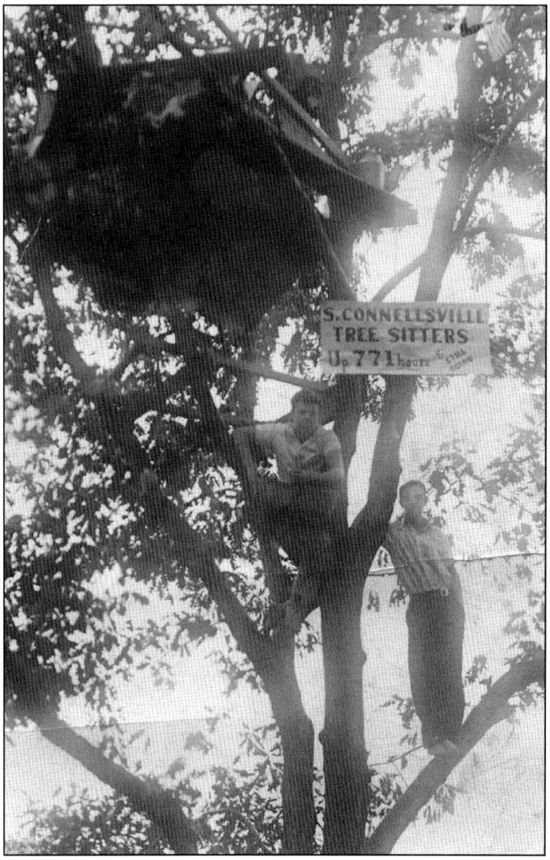

Donald Dasconi (left) and Harvey Shepley attempted to break the "world's" record for tree-sitting in South Connellsville. On the back of this photograph, they wrote, "Went up Friday, July 25, 1930. Will come down with record." Unfortunately, the outcome was not recorded. Sitting on poles and in trees were popular competitive activities for young people when funds were limited for other pastimes during the Great Depression.

The rabbits below, at Keagy's Drug Store, announced the arrival of spring in Connellsville. Every year, the front window was cleared for a display of Easter bunnies. Children delighted in visiting the window and trying to capture the attention of a rabbit. This photograph was taken in the mid-1950s. (Courtesy of Bill Sechler.)

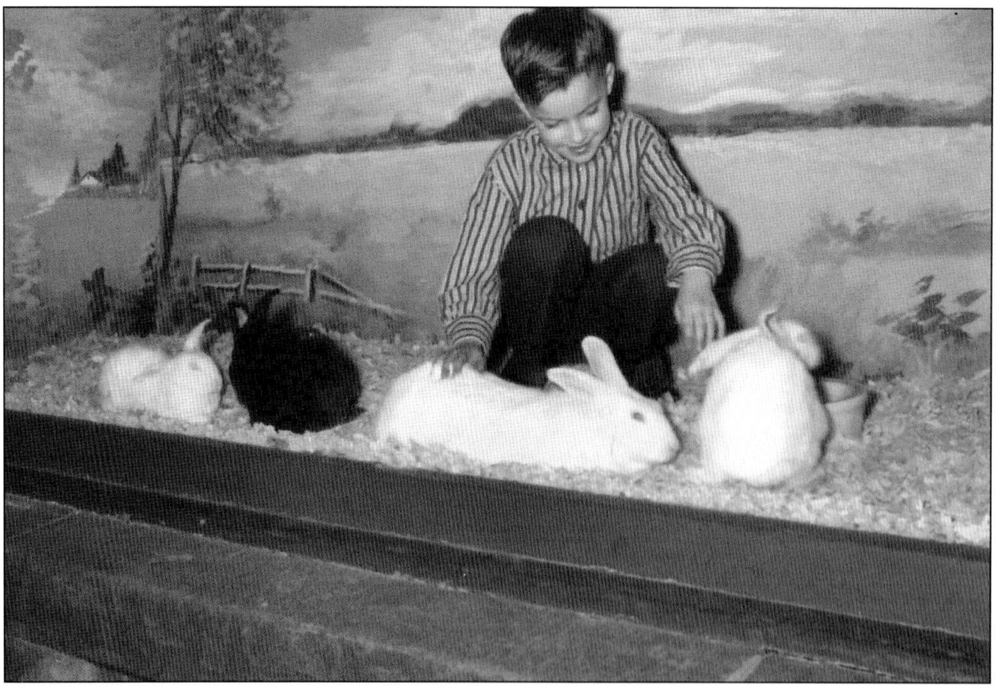

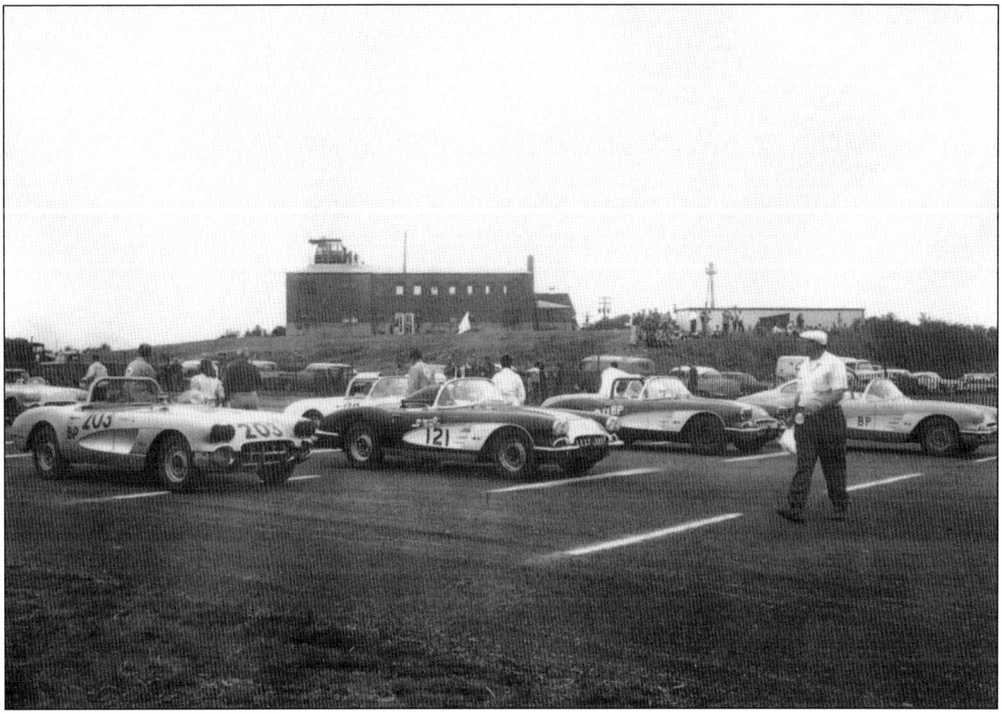

In addition to general aviation, Connellsville Airport served as a venue for Sports Car Club of America–sanctioned races seen above in 1960. While adults "played" at the airport, young people tested their design and driving skills in the soapbox derby below, held on Limestone Hill in the 1950s. The event was organized by the Greater Connellsville Jaycees. (Above, courtesy of Marion Scardina.)

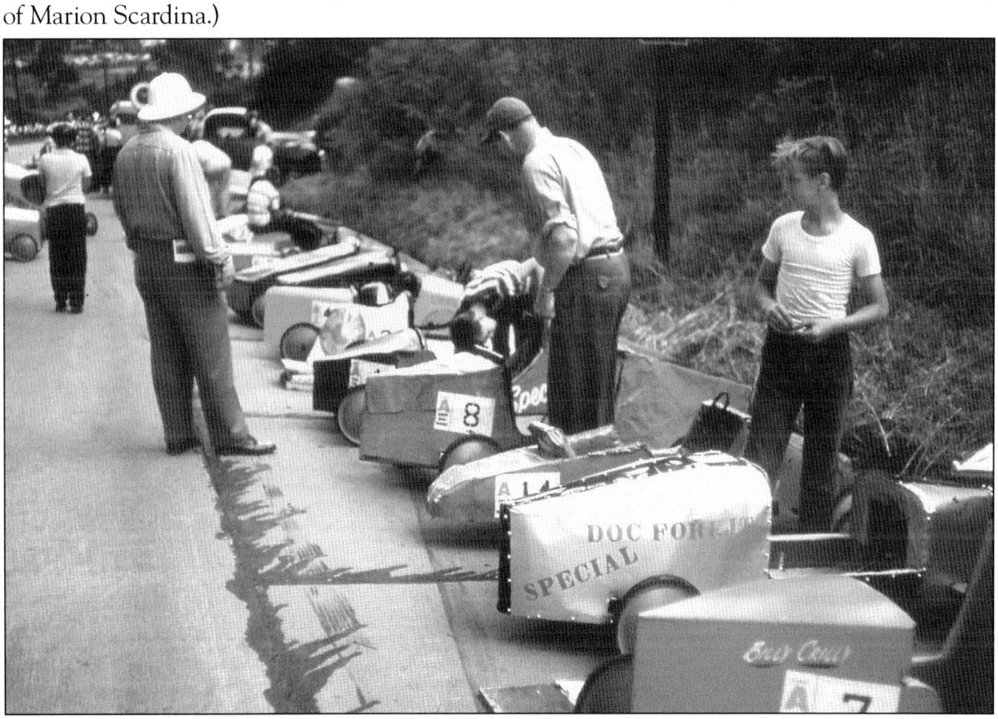

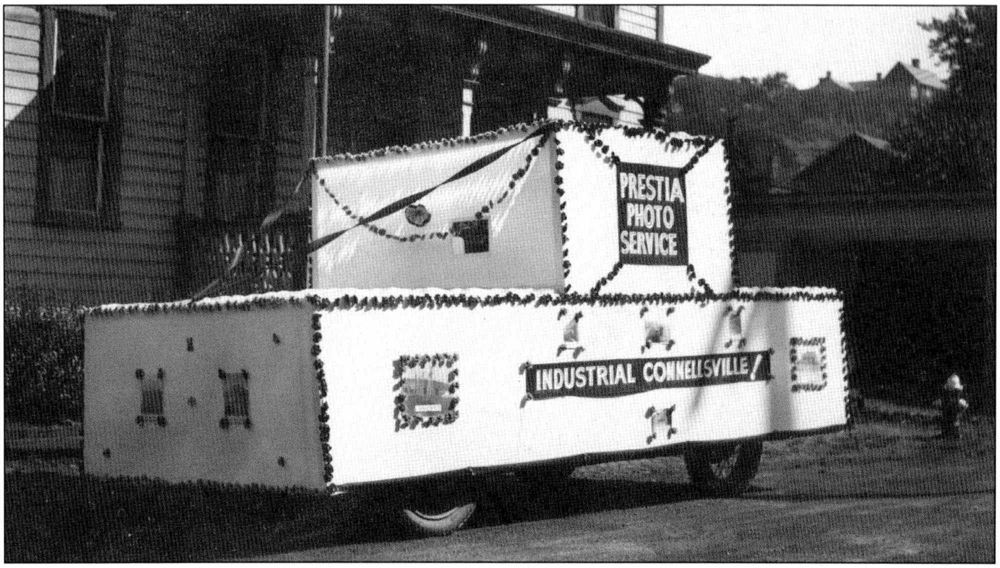

Felix Prestia Sr. operated a photographic service in the 1930s from a shop on West Apple Street. Seen here is a float Prestia entered in a Connellsville parade that celebrated local industries. In addition to developing pictures, Prestia was a commercial photographer for the B&O Railroad and local manufacturers. (Courtesy of Felix Prestia.)

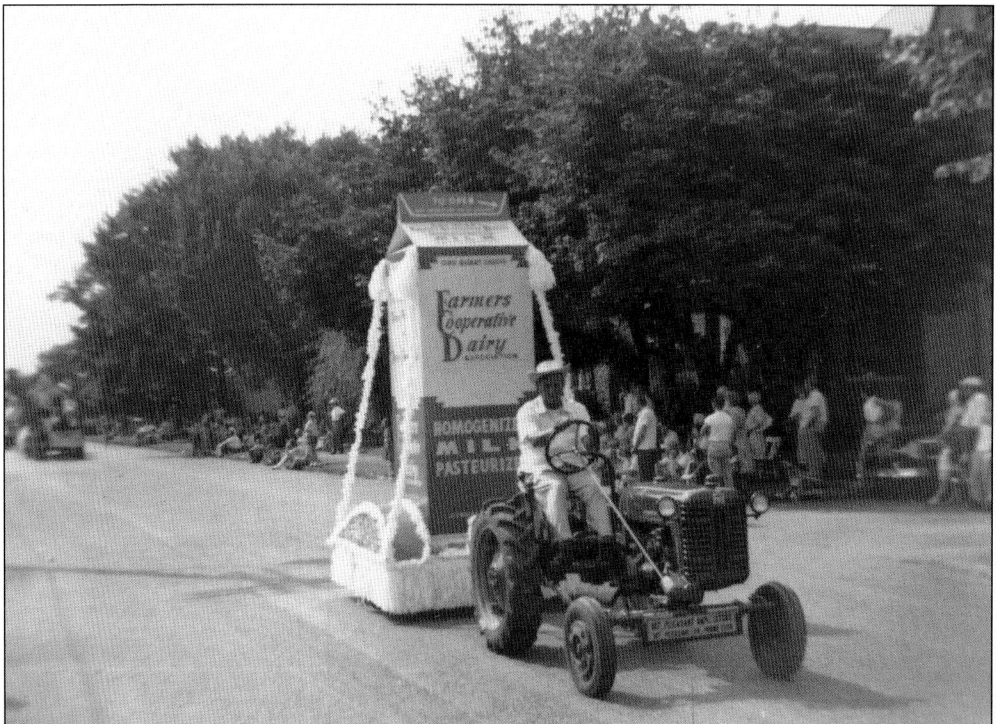

Connellsville inaugurated its sesquicentennial with a series of events, including a big parade on August 18, 1956, that highlighted industries and organizations. The Farmers Cooperative Dairy Association, located on Gallatin Avenue in Stahl's Square, made this entry representing their contribution to the city. (Courtesy of Tillie Berg.)

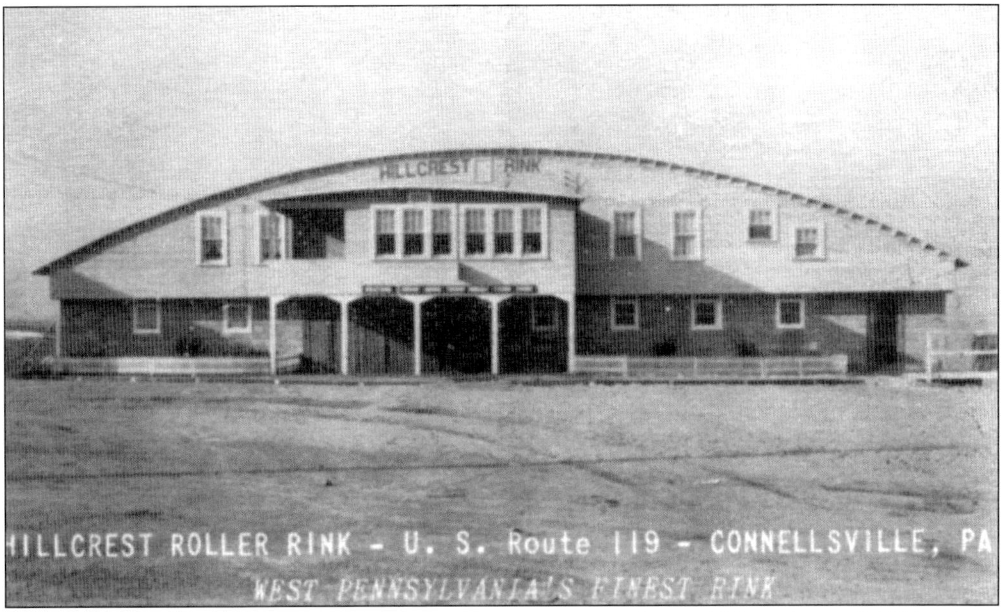

Hillcrest Roller Rink was a popular entertainment venue from 1930 to the mid-1950s. Area residents were urged to "skate for health's sake." The facility was on old Route 119 across from Dr. John E. Ellis's optometry office. For a period of time, the rink was owned by the Means family. (Courtesy of Margaret Molinaro.)

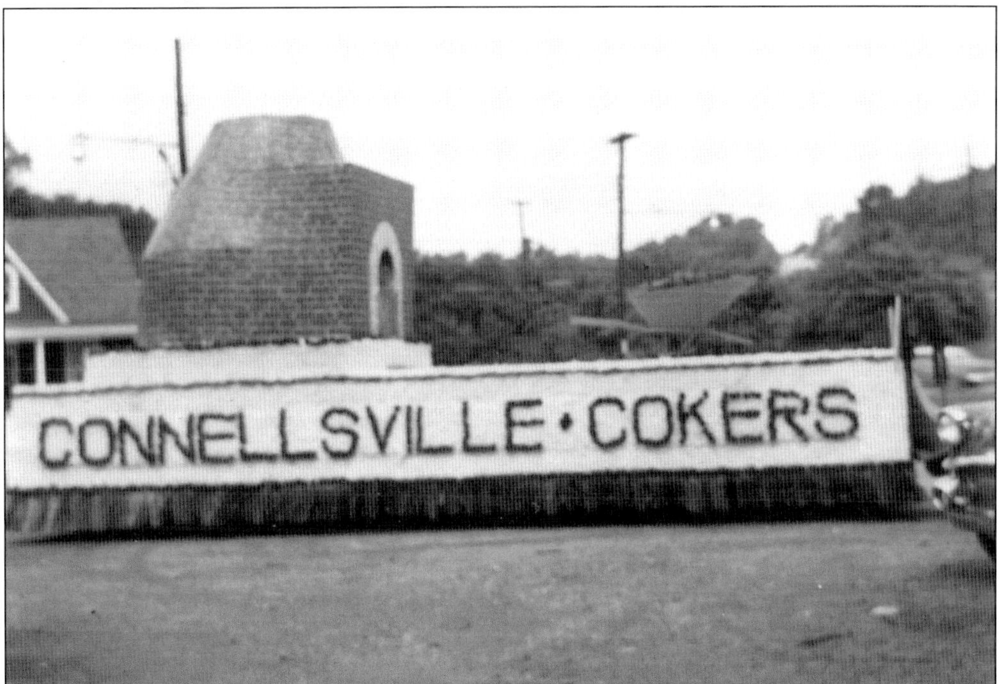

Connellsville celebrated the legacy of its largest industry and the name of its athletic teams with this replica of a coke oven built for the sesquicentennial parade in 1956. Connellsville High School teams were known as the Cokers until they became the Falcons after the district merger in 1966. (Courtesy of Sandy Russell.)

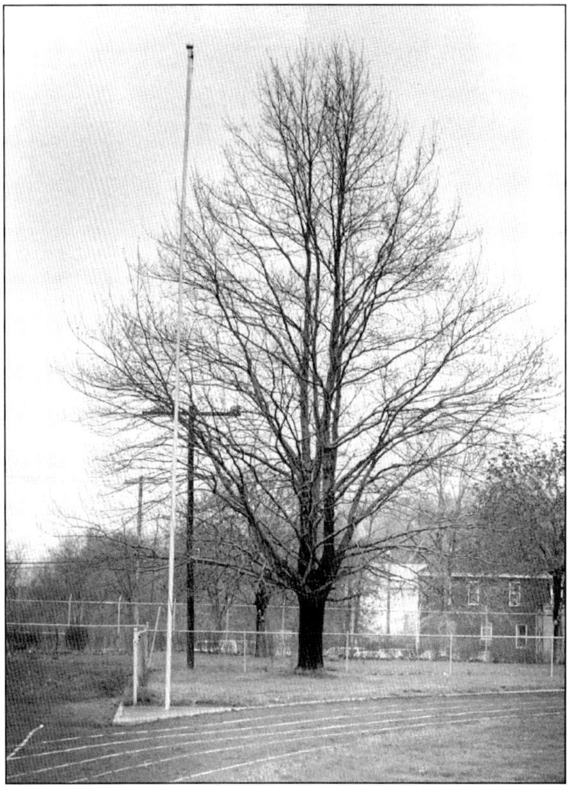

John Woodruff was given an oak tree sapling from the Black Forest in Germany after his winning performance in the 800 meters at the 1936 Berlin Olympics. The tree was originally planted on the lawn at the Carnegie Library but was subsequently moved to Campbell Field (Falcon Stadium). It continues to flourish as a symbol of Woodruff's outstanding victory and lifelong achievements.

In 1982, Connellsville citizens organized a five-kilometer race and related athletic events in honor of gold medal winner John Woodruff. The annual event is held in the second week in July and attracts over 500 participants from throughout the area. Woodruff is seen below on the far left starting the 1988 race at Falcon Stadium. (Courtesy of Paula Salatino Beucher.)

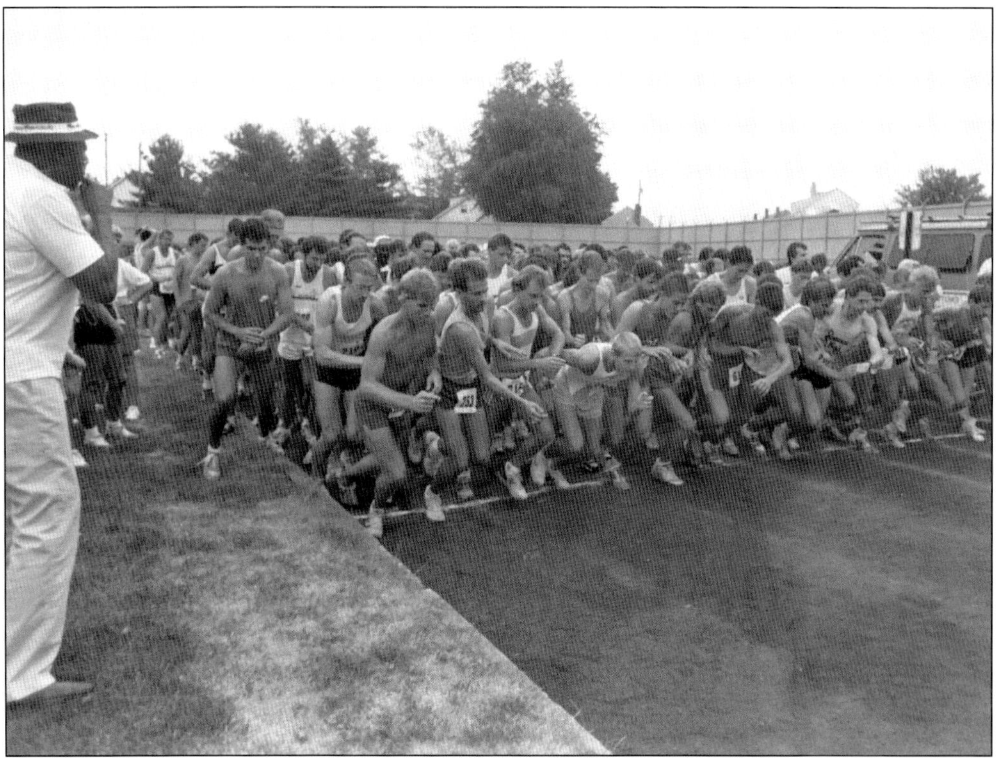

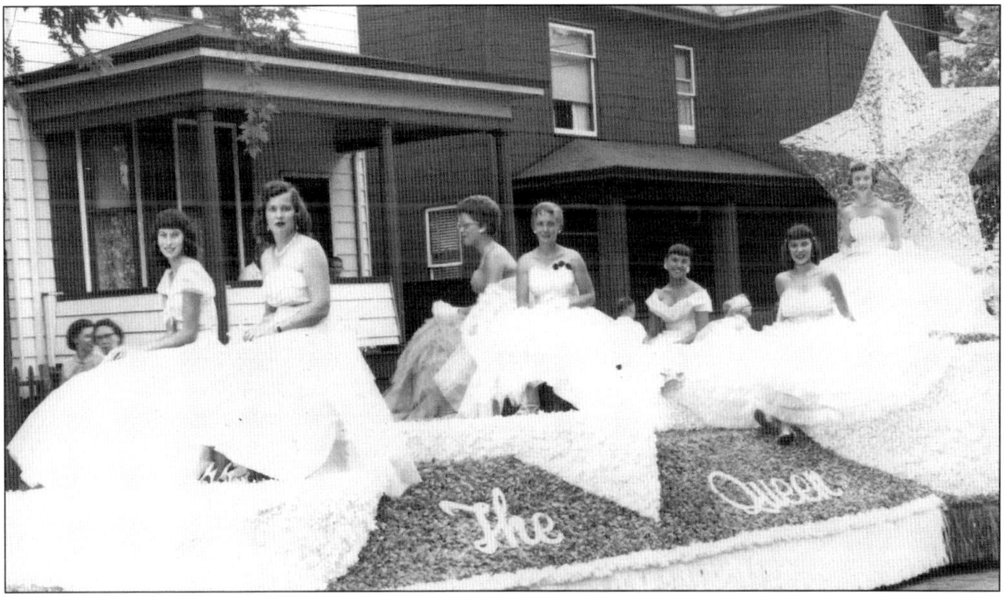

Seated in front of the star is Judy Swan Nardone, who reigned as queen of the Connellsville sesquicentennial. The members of her court were, from left to right, (first row) Rosaria Maricondi Findley and Jane Cavanaugh May; (second row) Mary Anne Molinaro Bock and Betty Rayger Markum; (third row) Carol Deleon Trimbath and Patti Graziano Toprani.

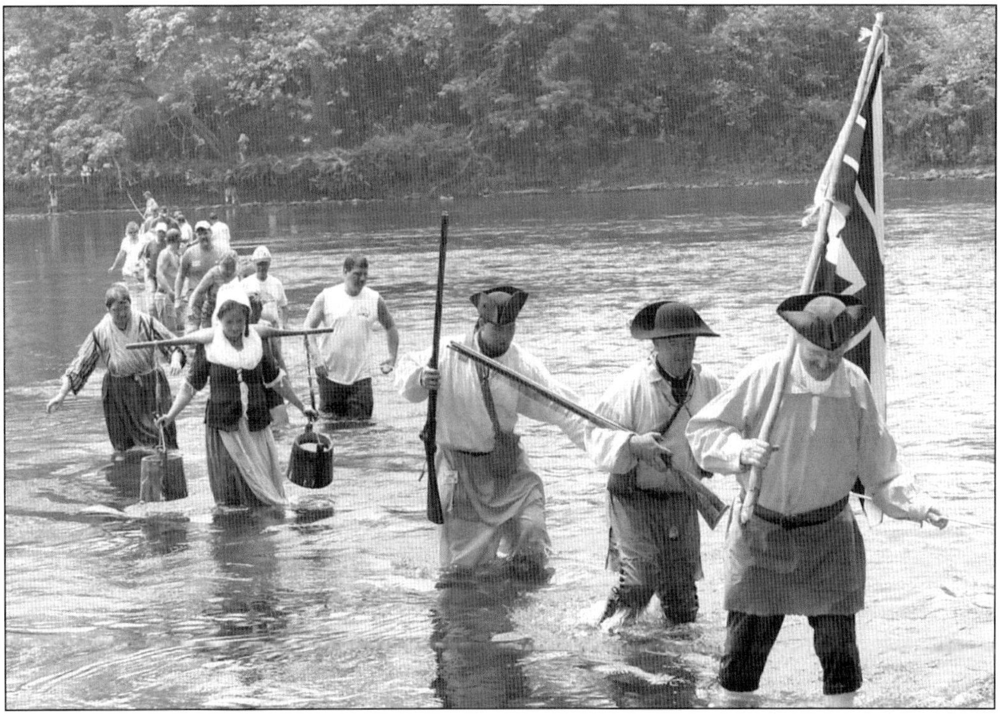

These men and women dressed in period clothes and others reenact Gen. Edward Braddock's historic crossing of the Youghiogheny River in Connellsville during the French and Indian War. In 2004, the Connellsville Area Historical Society began celebrating the crossing during the last full weekend in June. (Courtesy of Dexton Reed.)

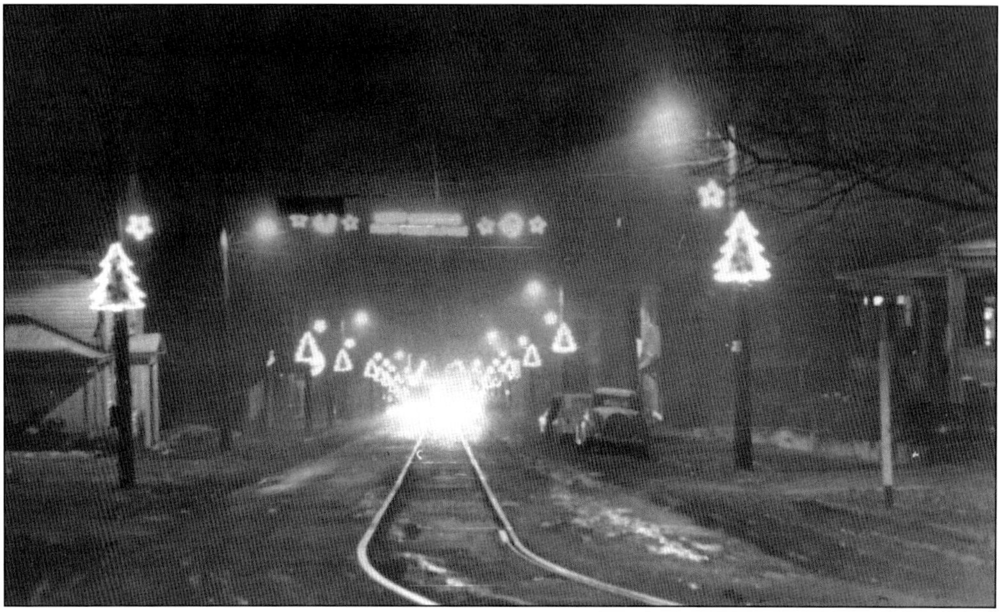

Connellsville knew how to make Christmas a special holiday with these green fluorescent lights, which thrilled residents and attracted shoppers to downtown stores. This nighttime scene from the early 1950s was taken at the top of the hill on East Crawford Avenue, looking west toward the central business district and the river.

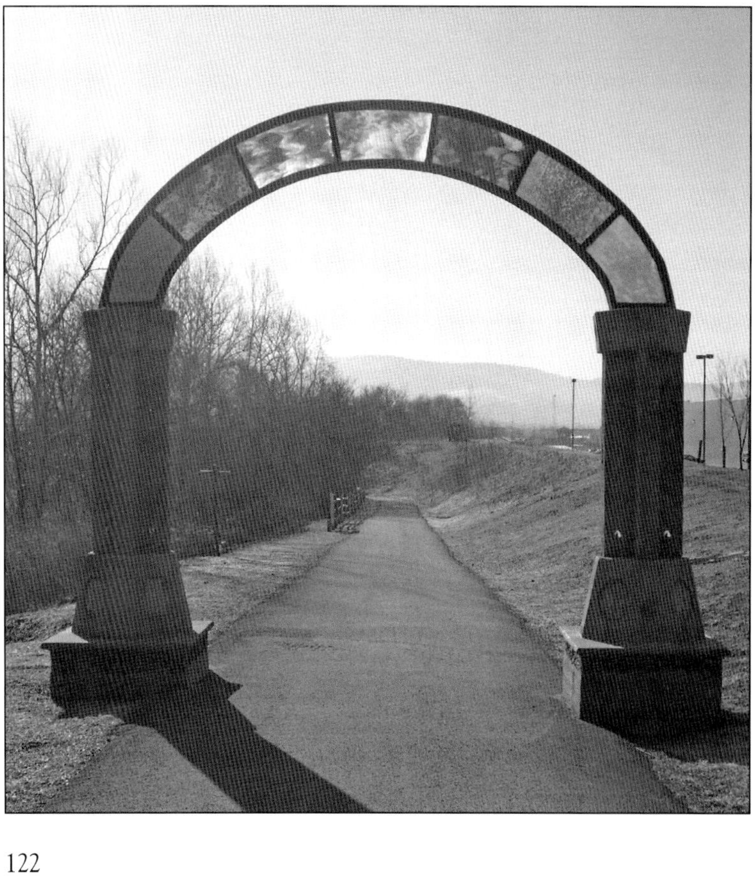

Steven Ficus designed this arch in 2008 to symbolize the arch that was made from coal and coke and erected on Brimstone Corner during the Connellsville centennial. He also added colorful glass panes at the top in recognition of the city's glass-making industry. The arch welcomes visitors to Connellsville at the northern trailhead of the Great Allegheny Passage, which is close to Yough River Park. (Courtesy of Daniel Cocks.)

Six

Representing a Hometown

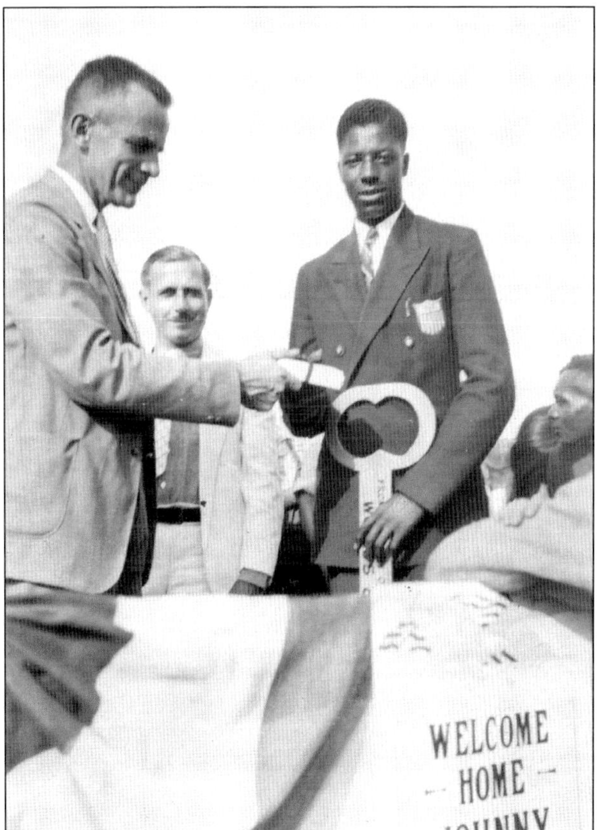

John Woodruff was born in South Connellsville and graduated from Connellsville High School in 1935. As a freshman at the University of Pittsburgh, Woodruff won the gold medal in the 800-meter run at the 1936 Olympics in Berlin, Germany. Adolph Hitler refused to acknowledge Woodruff's outstanding victory, but his hometown celebrated this singular achievement with a parade and by giving him the key to the city, as seen here. (Courtesy of Felix Prestia.)

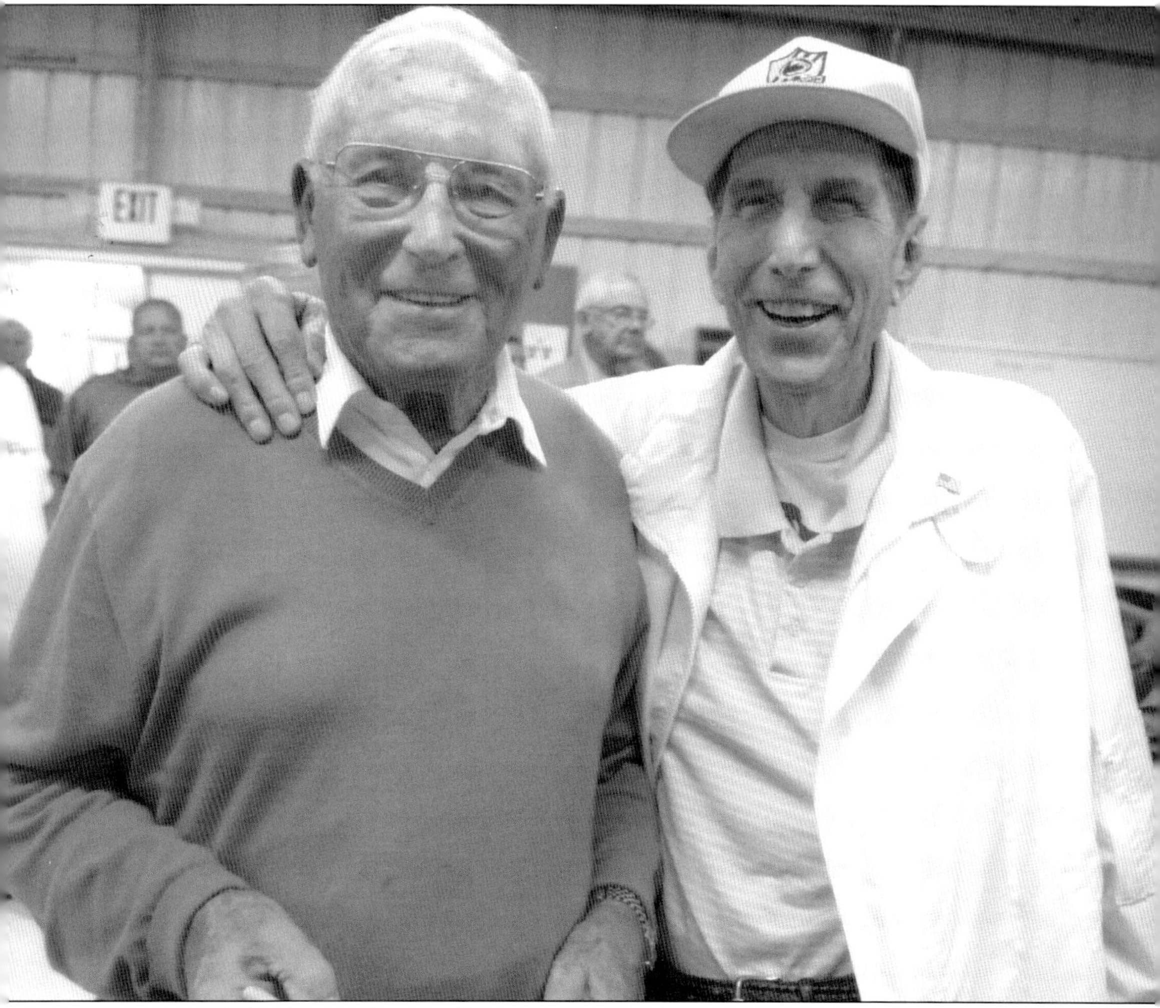

John Lujack (left) and John "Wally" Schroyer (right) were reunited in 2009 after Lujack was inducted into the Fayette County Sports Hall of Fame. Schroyer was subsequently inducted in 2012. The 1942 graduates of Connellsville High School were outstanding athletes in football, basketball, and track. Schroyer played fullback on the undefeated 1941 football team; Lujack played halfback. After one year of college, Lujack joined the Navy and saw action in the North Atlantic. His friend joined the Army, was wounded at Anzio, and was then held prisoner by the Germans for three years. After the war, Lujack returned to Notre Dame and led the Irish to three consecutive national titles, earning the Heisman Trophy in 1947. He played for the Chicago Bears and then went on to be a television commentator and a successful businessman. Wally Schroyer returned to Penn State, where he was the quarterback before the war; however, because of health issues related to his war injuries, he returned to Connellsville and entered local government. Schroyer served as the chief assessor and register of wills for Fayette County, retiring after 40 years of distinguished public service. (Courtesy of John "Wally" Schroyer.)

Florence Shutsy Reynolds learned to fly at the Connellsville Airport. After graduating from Dunbar High School, she joined the Women Airforce Service Pilots (WASP) in 1943 and completed training at Avenger Field in Sweetwater, Texas. As a WASP, she was one of the first women to fly military aircraft, releasing male pilots for combat missions. In 2010, Reynolds and other WASPs received the Congressional Gold Medal. (Courtesy of Tom Rusnack.)

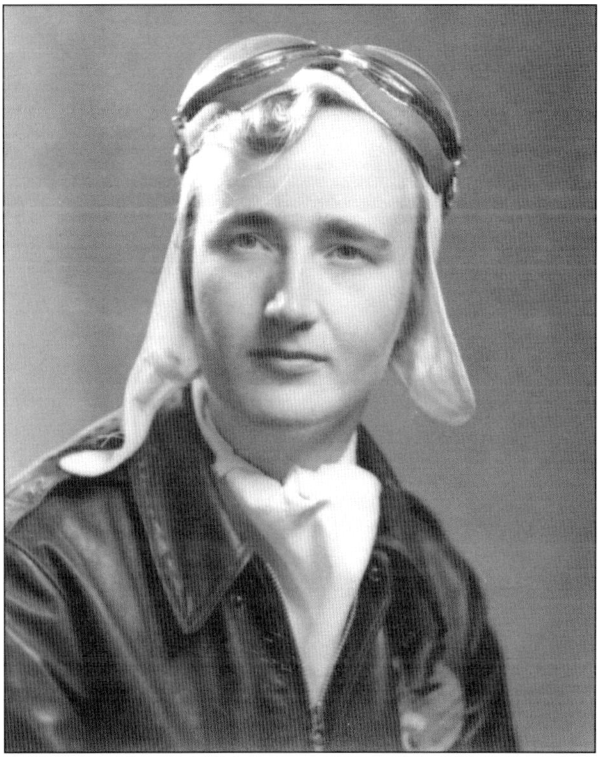

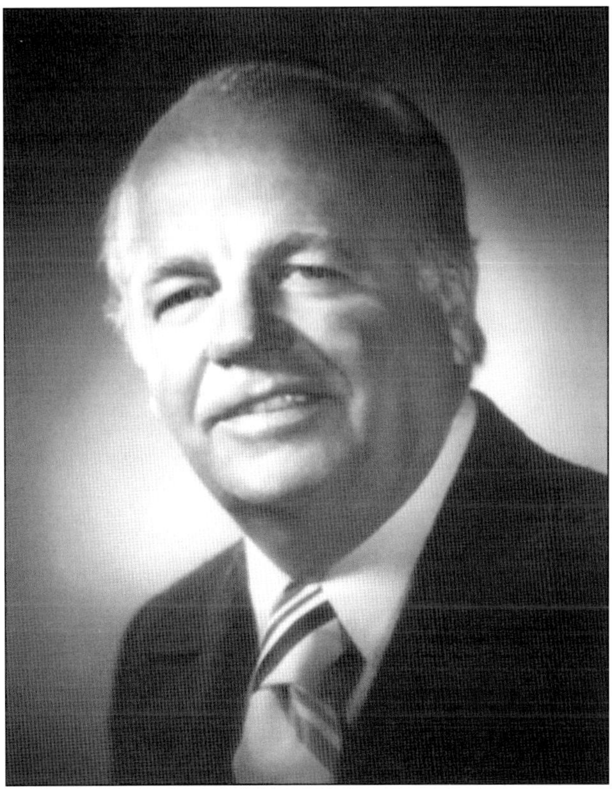

Ray Scott graduated from Connellsville High School in 1936. His career started in radio, and he then became the "Voice of the Green Bay Packers" for CBS television. He announced two Super Bowls as well as the famous NFL championship "ice bowl" game in 1967. Scott was named National Sportscaster of the Year on two occasions and was inducted into the Pro Football Hall of Fame. (Courtesy of Tom Rusnack.)

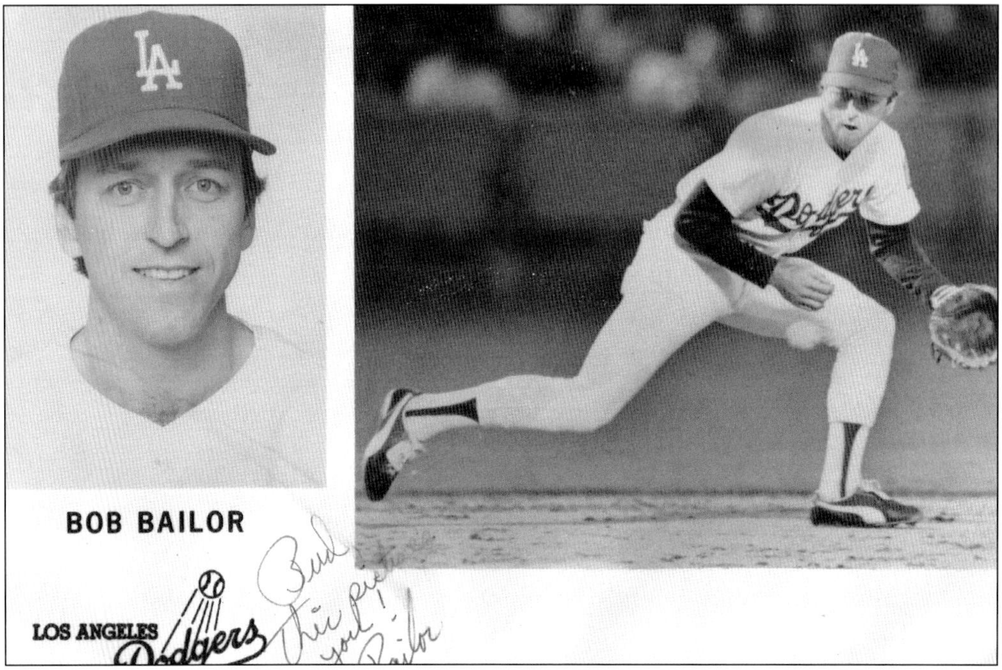

Bob Bailor grew up in Connellsville and graduated from Geibel Catholic High School in 1969. Bailor excelled in American Legion baseball and was drafted by the Baltimore Orioles after graduation. He played with Baltimore for six years, followed by stints with the Toronto Blue Jays, the New York Mets, and the Los Angeles Dodgers. After his release from the Dodgers in 1986, he coached for the Blue Jays. (Courtesy of Bud Murphy's.)

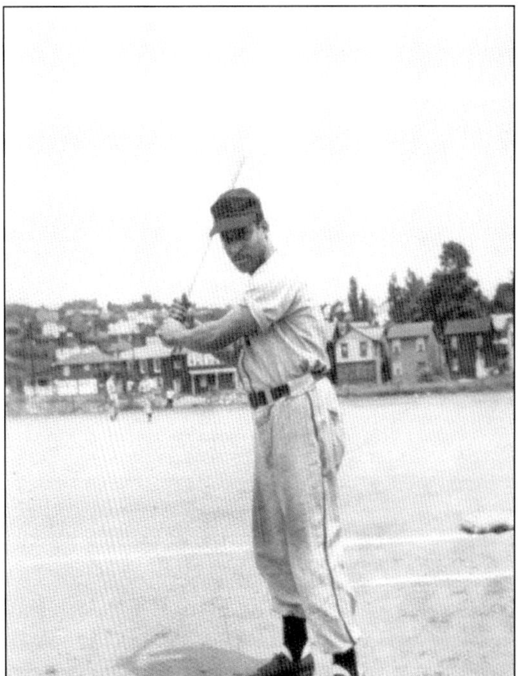

Connellsville native Cecil Cole was an outstanding pitcher in the local WPA-YMCA Junior Baseball League in the late 1930s. After serving in World War II, he played for one season with the Newark (New Jersey) Eagles, winning the Negro League World Championship in 1946. Upon his return to Connellsville, he worked for the Connellsville Housing Authority and served as a part-time scout for the Pittsburgh Pirates. (Courtesy of Howard Barnhart.)

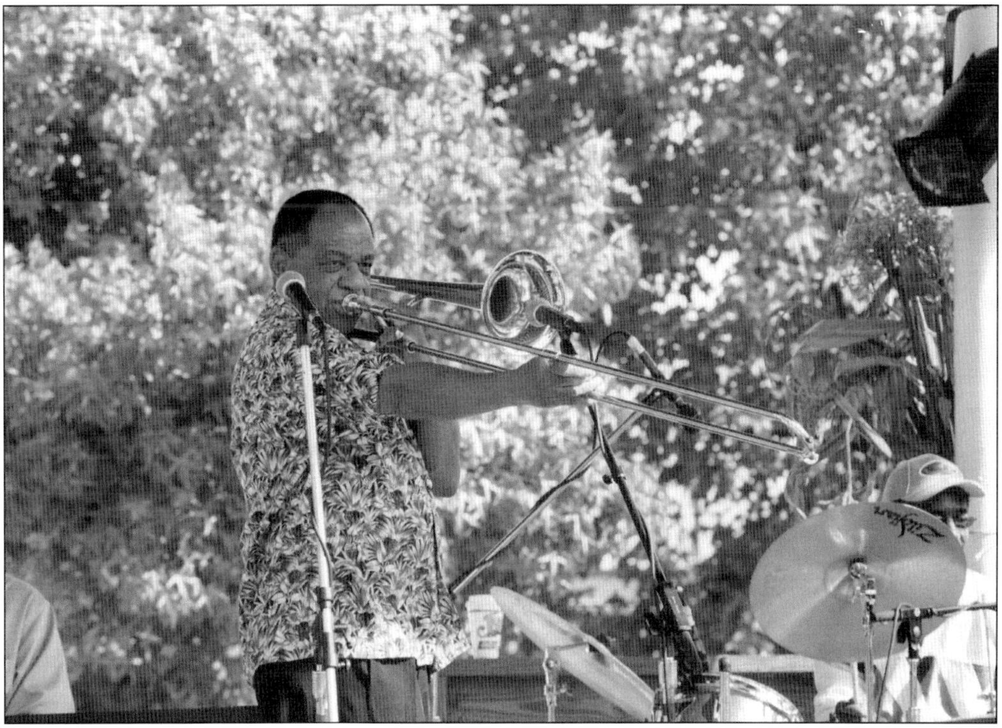

Harold Betters, better known as Mr. Trombone, is a Connellsville native and a distinguished jazz musician. During his career of over 50 years, he recorded several albums, appeared on television shows, and played with other notable artists such as Ray Charles, Louis Armstrong, and Al Hirt. He continues to entertain audiences throughout the country in clubs and at special events, including at Pittsburgh Steelers home games.

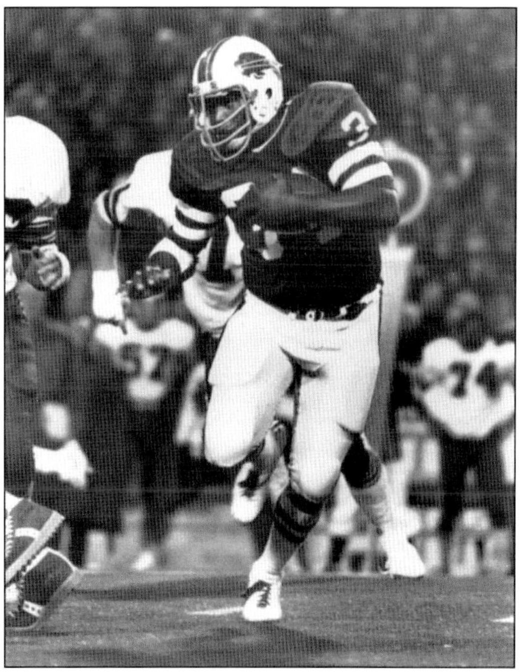

James Braxton was born in Vanderbilt and played basketball, football, baseball, and track at Dunbar High School. After the district merged with Connellsville, he continued to excel in these sports. Braxton attended West Virginia University (WVU), earning All America honors. After college, he played for the Buffalo Bills and the Miami Dolphins. He was inducted into the WVU Sports Hall of Fame in 1992 following his untimely death in 1986. (Courtesy of Pamela Braxton.)